Lecture Notes in Mathematics

A collection of informal reports and seminars
Edited by A. Dold, Heidelberg and B. Eckmann, Zürich

Series: Mathematics Institute, University of Warwick
Adviser: D. B. A. Epstein

141

Graham Jameson
University of Warwick, Coventry/England

T0220316

Ordered Linear Spaces

Springer-Verlag
Berlin · Heidelberg · New York 1970

PREFACE

Most of the familiar real linear spaces have a natural ordering. The study of ordered linear spaces was initiated in the late 1930's (by Riesz, Freudenthal, Kantorovič, Kakutani and others), since when a large and scattered research literature has grown up. This work is an attempt to provide a balanced introductory treatment of the various branches of the subject, without giving an encyclopaedic survey of any of them. Suggestions for further reading are distributed through the text. These do not constitute a complete bibliography of the topics considered, but it is hoped that they will be found sufficient to launch the reader into the relevant literature. The treatment of the elementary theory is quite detailed, because it is the author's belief that a good mastery of the basic material is a necessary condition for successful "advanced" work in any subject. The more elementary sections, in fact, formed the basis of an undergraduate course given at the University of Warwick in 1968.

Chapters 1 and 2 deal with the purely algebraic theory, while Chapters 3 and 4 are concerned with ordered topological linear spaces. Chapter 3 is independent of Chapter 2. A short final chapter on ordered algebras is designed to give a taste, rather than a systematic treatment, of this subject. However, this appears to be the natural context for certain results on monotonic linear mappings without which any account of ordered linear spaces would be incomplete.

An introductory Chapter 0 is included to summarise the terminology used concerning orderings and linear spaces (where alternatives exist), and the main results to be assumed known. The intention is purely to facilitate reference. Chapter 0 is neither self-contained (e.g. we do not repeat elementary definitions such as "topological linear space"), nor in any way a balanced summary of the subjects concerned.

No attempt is made to consider linear spaces over fields other than the real numbers, but some care is taken to point out which results require no scalar field at all, and therefore apply to ordered commutative groups. To preserve conceptual simplicity, however, we ignore any possible generalisations to non-commutative groups. For a detailed treatment of ordered groups, the reader is referred to Fuchs (1).

The choice of material is to some extent complementary to that in the book of Peressini (2). For example, more space is given to ordered normed spaces. The formulation and proof of 1.7.3 (possibly the most important theorem) are rather different from previous versions. A similar comment applies to the main theorem on extremal monotonic linear functionals (1.8.2), together with its topological version (3.1.12). Some of the material on cones with bases (sections 3.8 and 4.4.4) is new, at least in presentation.

The author is indebted to Prof. F. F. Bonsall, without whose constant advice and encouragement these notes would never have reached completion, and to Mrs Susan Elworthy for typing the text.

CONTENTS

CHAPTER 0: PRELIMINARIES

CHAPTER 1: ORDERED LINEAR SPACES

CHAPTER 2: LINEAR LATTICES AND RIESZ SPACES

CHAPTER 5: ORDERED ALGEBRAS

CHAPTER 0

PRELIMINARIES

"If and only if" is abbreviated everywhere to "iff". Set-theoretic inclusion is denoted by $\subseteq$, the symbol $\subset$ being reserved for proper inclusion. Set-theoretic difference is denoted by $\sim$. The set of positive integers is denoted by ω, the set of real numbers by R, and the set of non-negative real numbers by R^+. The closure of a set A is denoted by $\bar{A}$, and the interior by int A.

Results (and only results) are numbered within sections, those in section m.n being m.n.1, m.n.2, etc. The more important ones are dignified with the designation "theorem".

0.1. Orderings

The following (not entirely standard) terminology will be used. An <u>ordering</u> is a reflexive, transitive relation. Our usual notation for an ordering is $\leq$, and the following conventions will be used throughout: (a) $x \geq y$ means $y \leq x$, (b) $y < x$ (or $x > y$) means: $y \leq x$ and $y \not\geq x$. An ordering is <u>antisymmetric</u> if $x \leq y$ and $y \leq x$ implies $x = y$, and <u>connected</u> if for each pair x,y, either $x \leq y$ or $y \leq x$. A <u>total</u> ordering is a connected, antisymmetric ordering. If X,Y are ordered sets (the ordering in each set being denoted by $\leq$), then a mapping f from X to Y is <u>monotonic</u> if $f(x_1) \leq f(x_2)$ whenever $x_1 \leq x_2$.

If $a \leq b$, then the set $\{x : a \leq x \leq b\}$ will be denoted by $[a,b]$. Such sets are called <u>order-intervals</u>.

A subset A of an ordered set is <u>order-convex</u> if $a,b \in A$ and $a \leq x \leq b$ implies $x \in A$. The smallest order-convex set containing a given set A is

$$\{x : a \leq x \leq b \quad \text{for some } a,b \text{ in } A\}.$$

This set will be denoted by $[A]$, and called the <u>order-convex</u> <u>cover</u> of A.

A subset A of an ordered set is <u>directed upwards</u> (or <u>upward-directed</u>, depending on the needs of syntax) if, given two elements a,b of A, there is an element c of A such that $a \leq c$ and $b \leq c$. <u>Downward-directed</u> is defined similarly, and <u>directed</u> means "directed both upwards and downwards".

If $a \leq x$ for each a in A, we write $A \leq x$, and say that x is an <u>upper bound</u> of A. If A has an upper bound, it is said to be <u>majorised</u>. If $A \leq x$ and $x \in A$, then x is said to be a <u>greatest element</u> of A (such an element is unique if the ordering is antisymmetric). <u>Lower bounds</u> and <u>least elements</u> are defined similarly. If the set of upper (lower) bounds of A has a least (greatest) element, this is called a <u>supremum</u> (<u>infimum</u>) of A, and denoted by sup A (inf A).

An ordered set is said to be <u>order-complete</u> if every majorised set has a supremum. The following statement is equivalent: every minorised set has an infimum (consider the supremum of the set of lower bounds). An order-convex subset of an order-complete space is order-complete.

A <u>lattice</u> is a set with an antisymmetric ordering such that each pair of elements x,y has a supremum (denoted by $x \vee y$) and an infimum (denoted by $x \wedge y$). A lattice is <u>distributive</u> if

$$x \vee (y \wedge z) = (x \vee y) \wedge (x \vee z)$$

for all x,y,z. This is equivalent to the statement obtained by interchanging $\vee$ and $\wedge$. For an account of the theory of lattices, we refer to books on the subject; however, hardly any of this theory is required for our purposes.

<u>Interpolation properties</u>. For subsets A, B of an ordered set X, we write $A \leq B$ to mean $a \leq b$ for all a in A and b in B. Consider the following hierarchy of interpolation properties: if $A \leq B$, where

(i) A,B are non-empty subsets of X,

(ii) one of A,B is finite,

(iii) both A and B are finite,

then there exists x such that $A \leq x \leq B$. Condition (i) is equivalent to order-completeness (let B be the set of upper bounds of A). If condition (ii) holds, and the ordering is antisymmetric and directs X, then X is a lattice. Condition (iii) is called the Riesz interpolation property; it is easily seen by induction that it is sufficient for (iii) if the above condition holds for all two-point sets A,B.

0.2 Linear spaces

By a "linear space" we shall mean a linear space over the real field. The zero element will normally be denoted by 0. In this section and the next, we establish some notation for later use, and list the more important results that will be assumed known. The references for these, except where stated otherwise, are taken from Kelley-Namioka (1).

Let A be a subset of a linear space X. We denote by lin (A) and co(A) respectively the linear subspace generated by A and the convex cover of A. A point a_o is an internal (core) point of A if, given x in X, there exists $\delta > 0$ such that $a_o + \lambda x \in A$ whenever $|\lambda| \leq \delta$. We denote by A_o the set of internal points of A. If $0 \in A_o$, we say that A is absorbent (radial at 0).

0.2.1 (2.2, p. 15). If A is convex, $a \in A_o$, $b \in A$, and $0 < \lambda \leq 1$, then $\lambda a + (1 - \lambda)b \in A_o$.

A is said to be lineally closed if each line meets A in a closed subset of the line. Equivalently, if $x \notin A$, then x is an internal point of $X \sim A$. The lineal closure A_c of A is the complement of the set of internal points of $X \sim A$. Clearly, A is lineally closed iff $A_c = A$. If A is convex, then $x \in A_c$ iff there exists y in X such that $x + \lambda y \in A$ for $0 < \lambda < 1$ (note that if $x \in A$, this holds with $y = 0$).

An extreme point of a convex set A is a point b of A such that

$A \sim \{b\}$ is convex. An <u>extreme subset</u> (or <u>support</u>) of A is a convex subset B such that the following holds: if $a, a' \in A$, $0 < \lambda < 1$ and $\lambda a + (1-\lambda)a' \in B$, then $a, a' \in B$.

We denote by X' the space of all linear functionals on X. If (X,Y) is a separated dual pair of linear spaces, we regard elements of Y as linear functionals on X (and conversely), and if $A \subseteq X$, we write

$$A^o = \{y \in Y : y(a) \geq -1 \text{ for all } a \text{ in } A\}.$$

If $B \subseteq Y$, we define B^o similarly as a subset of X. If A is symmetric (i.e. if $A = -A$), then A^o is the set of y in Y such that $|y(a)| \leq 1$ for all a in A, and if A admits multiplication by positive scalars, then A^o is the set of y in Y such that $y(a) \geq 0$ for all a in A.

A functional (i.e. a real-valued function) p on X is <u>sublinear</u> if $p(\lambda x) = \lambda p(x)$ $(x \in X, \lambda > 0)$ and $p(x + y) \leq p(x) + p(y)$ $(x, y \in X)$.

<u>0.2.2</u> (Hahn-Banach; 3.4, p.21). Let X be a linear space, and let p be a sublinear functional on X. Suppose that f is a linear functional defined on a linear subspace E of X, and satisfying $f(y) \leq p(y)$ $(y \in E)$. Then there is a linear functional $\overline{f}$ on X such that $\overline{f}(y) = f(y)$ $(y \in E)$ and $\overline{f}(x) \leq p(x)$ $(x \in X)$.

<u>0.2.3 COROLLARY</u>. Let X be a linear space, and let p be a sublinear functional on X. Given x_0 in X, there exists f in X' such that $f(x) \leq p(x)$ $(x \in X)$ and $f(x_0) = p(x_0)$.

<u>0.2.4</u> (Eidelheit; 3.8, p.22). Suppose that A,B are convex subsets of a linear space X such that $A_o \neq \emptyset$ and $A_o \cap B = \emptyset$. Then there is a non-zero element f of X' such that $\inf f(A) \geq \sup f(B)$.

The following notation will be used consistently to denote certain spaces of sequences:

s the space of all real sequences;

m the space of all bounded real sequences;

c_o the space of all real sequences that converge to 0;

l_1 the space of all real sequences $\{\xi_n\}$ such that $\sum_{n=1}^{\infty} |\xi_n|$ is convergent;

F the space of all real sequences having only a finite number of non-zero terms.

Generally, a sequence x is denoted by $\{\xi_n\}$. When a sequence in F is denoted by $(\xi_1, \ldots, \xi_n)$, it is implied that $\xi_r = 0$ for $r > n$. When there is no danger of confusion, we denote by e the sequence having every term equal to 1, and by e_n the sequence having 1 in place n and 0 elsewhere.

0.3 Topological linear spaces

Let X be a topological linear space (or commutative group). When no confusion can arise owing to the presence of different topologies, we denote by Ⓝ(X) the family of all neighbourhoods of 0. By a <u>local base</u> we mean a base of neighbourhoods of 0. We denote by X* the space of all continuous linear functionals on X. If (X,Y) is a dual pair, $\sigma(Y)$ denotes the weak topology induced by Y on X and its subsets.

<u>0.3.1</u> (5.2, p.35). If A and B are compact, then so is A + B. If A is compact and B is closed, then A + B is closed.

<u>0.3.2</u> (6.7, p. 48). A compatible topology is pseudo-metrizable iff there is a countable local base. The topology can then be given by a pseudo-metric d such that $d(x + z, y + z) = d(x,y)$ for all x,y,z (i.e. an invariant pseudo-metric).

<u>0.3.3</u>. If τ is a locally convex topology for a linear space X, then:

(i) (17.1, p.154) Convex subsets of X have the same closure with respect to τ and $\sigma(X^*)$.

(ii) (17.5, p. 155) A subset A of X is τ-bounded iff it is

$\sigma(X^*)$-bounded, and this occurs iff $f(A)$ is bounded for each f in X^*.

0.3.4 (Day (1), p.19). If X is a linear space, and p is a real-valued function on X, then $\{f \in X' : f(x) \leq p(x)$ for x in $X\}$ is compact with respect to $\sigma(X)$.

A real locally convex space X is said to be bornological if each convex set that absorbs all bounded sets is a neighbourhood of 0, and barrelled if each closed, convex absorbent set is a neighbourhood of 0. Every locally convex, metrizable space is bornological, and every Fréchet space (i.e. complete, metrizable, locally convex space) is barrelled. X is said to be a Mackey space if $B^o \in \mathfrak{N}(X)$ whenever B is a convex, $\sigma(X)$-compact subset of X^*. Every bornological or barrelled space is a Mackey space.

The next group of results concerns extreme points.

0.3.5 (Krein-Mil'man: 15.1, p. 131). Let A be a compact convex subset of a locally convex, Hausdorff space. Then:

(i) Each closed extreme subset of A contains an extreme point of A.

(ii) A is the closed, convex cover of the set of its extreme points.

The following is a purely algebraic corollary of 0.3.5 and 0.3.4:

0.3.6. Let p be a sublinear functional on a linear space X, and let

$$H = \{f \in X' : f(x) \leq p(x) \quad \text{for all } x \text{ in } X\}.$$

Then, given x_o in X, there exists f in X' such that $f(x_o) = p(x_o)$ and f is an extreme point of H.

In the following, $\overline{co}(A)$ denotes the closure of the convex cover of A.

0.3.7. (Mil'man; 15.2, p. 132). If A is a closed subset of a locally convex, Hausdorff space, and $\overline{co}(A)$ is compact, then every extreme point of $\overline{co}(A)$ is in A.

0.3.8 (Schaefer (4), 1.6, p. 18). Suppose that τ_1, τ_2 are Hausdorff topologies for a linear space X, that τ_2 is smaller than τ_1, and that τ_1 has a local base consisting of τ_2-complete sets. Then τ_1 is complete.

0.3.9 (Krein-Šmul'jan; 22.6, p.212). Let X be a Fréchet space, and let A be a convex subset of X*. If Ⓑ is a local base in X, and $A \cap U^o$ is a σ(X)-closed for each U in Ⓑ, then A is σ(X)-closed.

We shall require Ascoli's theorem in the following form:

0.3.10 (cf. Kelley (1), p.234). Let X,Y be normed linear spaces, and let F be a bounded set of linear mappings from X to Y. Suppose that E is a totally bounded subset of X, and that $\{f(x) : f \in F\}$ is totally bounded for each x in E. Then, for each $\varepsilon > 0$, there is a finite subset $\{f_1, \ldots, f_n\}$ of F such that, given f in F, there exists i such that

$$\|f(x) - f_i(x)\| < \varepsilon \ (x \in E).$$

In the next two results, B(S) denotes the space of all bounded real-valued functions on a set S, and, when S is a topological space, C(S) denotes the space of all continuous real-valued functions on S. When S is compact, C(S) is contained in B(S). We assume B(S) to be topologised by the supremum norm.

0.3.11 (Kakutani; Simmons (1), p. 158). Let S be a compact topological space, and let A be a linear sublattice of C(S). Suppose that, given distinct points s,t of S, there exists g in A such that $g(s) \neq 0$ and $g(t) = 0$. Then A is dense in C(S).

The next result enables us to apply 0.3.11 to algebras of functions.

0.3.12. (Stone; Simmons (1), p. 159). For any set S, a closed subalgebra of B(S) is a sublattice.

CHAPTER 1

ORDERED LINEAR SPACES

This chapter is concerned with the purely algebraic theory of ordered linear spaces with no extra structure assumed. The basic correspondence between wedges and orderings in a real linear space is established in Section 1.1. Briefly, it is natural to consider orderings that are invariant under translation and positive scalar multiplication, and such an ordering is completely specified by the set of elements x such that $x \geq 0$. A similar construction applies in commutative groups if all mention of scalar multiplication is left out.

This introduces a duality which persists throughout the theory: statements can be expressed either in order-theoretic terminology or in terms of the positive wedge. Usually, the motivation for our work comes from order-theoretic concepts.

Sections 1.2, 1.3 and 1.4 are quite elementary. The ideas introduced in 1.3 are basic for some of our later work, and give a good illustration of the duality just mentioned: in linear space terminology, an order-unit is simply an internal point of the positive wedge P, and an ordering is Archimedean iff P is lineally closed.

Sections 1.5, to 1.8 are concerned with linear mappings and functionals. 1.5 is introductory. In 1.6 we show how the basic theorem (1.6.1) on extension of monotonic linear functionals is connected with the Hahn-Banach theorem: either can be deduced quite easily from the other. In 1.7 we tackle one of the most basic questions about ordered linear spaces: when can a linear functional be expressed as the difference between two monotonic linear functionals? The problem can be extended by requiring the functionals to be bounded on a given absorbent set; by doing so, we make our answers immediately applicable to the topological case, when continuity is required of the functionals (see Chapter 3). In 1.8 we show how to characterise extremal elements of the wedge of monotonic linear functionals, and give a Hahn-Banach-like theorem on the existence of such elements. This result, too, is fundamental for parts of our later theory.

The significance of bases for cones, considered in 1.9, will become more apparent in Chapter 3. Section 1.10 is not required for later chapters, but is of some interest in its own right.

1.1. Wedges, cones and orderings

Let X be a real linear space. A non-empty subset P of X will be called a <u>wedge</u> if the following two conditions hold:

$$P + P \subseteq P,$$

$$\lambda P \subseteq P \text{ for } \lambda \geq 0.$$

The ordering of X associated with a wedge P is the relation $\leq$ defined by:

$$x \leq y \iff y - x \in P.$$

It is clear that this is a reflexive, transitive relation connected with the linear structure of X by the conditions:

$$x \leq y \text{ implies } x + z \leq y + z \text{ for all } z \text{ in } X \quad (1),$$

$$x \leq y, \lambda \geq 0 \text{ implies } \lambda x \leq \lambda y. \quad (2)$$

By a <u>linear</u> ordering of a real linear space we shall mean an ordering satisfying conditions (1) and (2), and by an <u>ordered linear space</u> we shall mean a real linear space with a linear ordering. We shall adopt the convention that whenever we speak of an ordering of a linear space, we shall mean a linear ordering.

<u>1.1.1</u>. Let $(x,\leq)$ be an ordered linear space, and let $P = \{x : x \geq 0\}$. Then P is a wedge, and $\leq$ is the ordering associated with P.

<u>Proof</u>. If $x,y \in P$, then $x + y \geq x$, by (1), so $x + y \geq 0$, by transitivity. Hence $x + y \in P$. If $x \in P$ and $\lambda \geq 0$, then $\lambda x \in P$, by (2). Hence P is a wedge. By (1), we have

$$x \leq y \iff y - x \geq 0 \iff y-x \in P,$$

so that $\leq$ is the ordering associated with P.

Hence there is a one-to-one correspondence between linear orderings of X and wedges in X, and the study of ordered linear spaces is, in a sense, simply the study of wedges. However, it will be seen that the motivation for most of our work comes from the concepts associated with orderings.

Similarly, we define an ordered commutative group to be a commutative group with an ordering satisfying (1). Such orderings correspond in the above way to semigroups containing 0. Quite a lot of our elementary theory applies to ordered commutative groups.

We shall consistently use the letter P to denote the positive wedge (or semigroup). The next three results show how basic properties of the ordering correspond to properties of P.

1.1.2. The ordering is antisymmetric iff $P \cap (-P) = \{0\}$.

Proof. Necessity is obvious. Suppose that $P \cap (-P) = \{0\}$ If $x \leq y$ and $y \leq x$, then $y - x \in P \cap (-P)$, so $y = x$.

A wedge P such that $P \cap (-P) = \{0\}$ will be called a cone. (Some writers use the term cone where we use wedge and strict cone where we use cone.) For any wedge P, it is clear that $P \cap (-P)$ is a linear subspace.

1.1.3. The following statements are equivalent:

(i) $P - P = X$.

(ii) Given x in X, there exists u in X such that $u \geq x$ and $u \geq 0$.

(iii) X is directed by the ordering.

Proof. (i) => (ii). Given x in X, there exist u,v in P such that $x = u - v$. Then $u \geq x$ and $u \geq 0$.

(ii) => (iii). Take x,y in X. There exist u,v in P such that $u \geq x$ and $v \geq y$. Then $u + v \geq x$ and $u + v \geq y$. Hence X is directed upwards. This clearly implies that it is also directed downwards.

(iii) => (i). Take x in X. There exists u in X such that $u \geq x$ and $u \geq 0$. Let $v = u - x$. Then $u,v \in P$ and $x = u - v$.

We say that P <u>generates</u> X if $P - P = X$. In any case, it is clear that $P - P$ is the linear subspace (or subgroup) generated by P.

<u>1.1.4</u>. The ordering is connected iff $P \cup (-P) = X$.

<u>Proof</u>. Clear.

<u>Some elementary properties</u>

We notice that $\{x : x \geq a\} = a + P$, and that $[a,b] = (a + P) \cap (b - P)$.

<u>1.1.5</u>. (i) If $a_i \leq b_i$ $(i = 1,2)$, then $a_1 + a_2 \leq b_1 + b_2$.

(ii) $[a_1,b_1] + [a_2,b_2] \subseteq [a_1 + a_2, b_1 + b_2]$.

<u>Proof</u>. (i) We have $a_1 + b_1 \leq a_1 + b_2 \leq a_2 + b_2$. (ii) follows.

The following algebraic properties of wedges are elementary:

(i) A wedge is convex.

(ii) If P,Q are wedges, then so is P + Q. If P,Q are cones and $P \cap (-Q) = \{0\}$, then P + Q is a cone.

(iii) The image and inverse image of a wedge under a linear mapping are wedges.

(iv) The intersection of a family of wedges is a wedge. The smallest wedge containing a set A is the set, denoted by pos A, of positive linear combinations of elements of A. pos A is a cone iff $\lambda_1 a_1 + \ldots + \lambda_n a_n = 0$, where $a_i \in A \sim \{0\}$ and $\lambda_i \geq 0$, implies that each $\lambda_i = 0$.

<u>1.1.6.</u> If A is convex, and $P = \bigcup\{\lambda A : \lambda > 0\}$, then P is a wedge. If $0 \notin A$, then P is a cone.

<u>Proof</u>. Clearly, $\lambda P = P$ for $\lambda > 0$. If $a,b \in A$ and $\lambda, \mu > 0$,

then $\lambda a + \mu b = (\lambda + \mu)c$, where $c \in A$. Hence P admits addition. If P is not a cone, then there exist a, b in A and $\lambda, \mu > 0$ such that $\lambda a = -\mu b$. Then $0 = (\lambda + \mu)^{-1} (\lambda a + \mu b) \in A$.

If E is a subgroup of an ordered commutative group X, then the restriction to E of the ordering in X is clearly the ordering of E associated with the semigroup $E \cap P$.

We mention one point specifically concerning groups. If an element x of P has finite order n, then $-x = (n-1)x \in P$. If each element of the group has finite order, it follows that P is a subgroup, and that the "ordering" is simply an equivalence relation. So to be of much interest from the order-theoretic point of view, a group must contain elements of infinite order.

Examples

(i) The usual ordering of R is that associated with the cone R^+. The only wedges in R are R^+, R^-, R, $\{0\}$.

(ii) Let X be a linear space of real-valued functions on a set S. The "natural" (or "usual") ordering of X is given by:

$$x \leq y \text{ iff } x(s) \leq y(s) \text{ for all } s \text{ in } S.$$

This is the ordering associated with the cone of functions whose values are non-negative.

(iii) Let X be a linear space of real sequences (e.g. s, m, c_0, l_1, F). The "usual" ordering of X is defined as in (ii). Some other orderings are those associated with the following wedges:

(a) Let P_s be the set of sequences in X having all partial sums non-negative. This is a cone which (in any of the spaces listed above) properly contains the usual cone.

(b) Let P_d be the set of decreasing (i.e. non-increasing) non-negative sequences in X. (The notations P_s and

P_d will be used consistently in the sense defined here.)

(c) (The <u>lexicographic</u> ordering.) Let P be the set of sequences in X whose first non-zero term is positive, together with 0. Then P is a cone giving a total ordering of X.

(iv) R^2. The orderings described in (iii) can be applied to R^2. The positive cone for the lexicographic ordering is the right half-plane, including the upper half of the vertical axis. Any wedge in R^2 consists of the region between two half-lines, which may coincide, and may or may not be included. A half-plane including its bounding line is a wedge, but not a cone.

(v) The set of all real polynomials forms a linear space, isomorphic to the space F of all finite sequences. Let the space of polynomials have the natural ordering obtained by regarding its elements as functions on $[0,1]$. Then the positive cone contains the isomorphic image of P_s (see (iiia)), since

$$a_o + a_1x + \dots + a_nx^n = (1 - x)(A_o + A_1x + \dots + A_{n-1}x^{n-1}) + A_nx^n,$$

where

$$A_r = a_o + a_1 + \dots + a_r \quad (1 \leq r \leq n).$$

Inclusion is proper, since $(1 - x)^2$ corresponds to $(1,-2,1)$, which is not in P_s.

<u>Suprema and infima</u>

We notice that a_o is an upper bound of A iff $A \subseteq a_o - P$, that a_o is a maximal element of A iff $(a_o + P) \cap A = a_o$, and that a_o is a supremum of A iff $a_o + P = \bigcap\{a + P : a \in A\}$.

<u>1.1.7</u>. Let A,B be subsets of an ordered commutative group X. If

A,B have suprema a_0, b_0, then $a_0 + b_0$ is a supremum of A + B. If A,B have infima a_1, b_1, then $a_1 + b_1$ is an infimum of A + B.

Proof. Clearly, $a_0 + b_0 \geqslant A + B$. Suppose that $x \geqslant A + B$, and take b in B. Then $x - b \geqslant A$, so $x - b \geqslant a_0$. Hence $x - a_0 \geqslant B$, so $x - a_0 \geqslant b_0$. The case of infima is similar.

1.1.8. If $a_0 = \sup A$, then $-a_0 = \inf(-A)$. If X is a linear space, and $\lambda > 0$, then $\lambda a_0 = \sup(\lambda A)$.

Proof. Immediate.

Directed subsets

By 1.1.3, a subgroup E is directed iff $E = E \cap P - E \cap P$.

Some elementary, but useful, properties of upward-directed sets are summarised in the next theorem.

1.1.9. Let A be an upward-directed subset of an ordered linear space. Then:

(i) If B is also upward-directed, then so is A + B.

(ii) co(A) is upward-directed.

(iii) A - P is convex.

(iv) If f,g are monotonic, real-valued functions on X that are bounded above on A, then

$$\sup(f + g)(A) = \sup f(A) + \sup g(A).$$

Proof. (i) Easy.

(ii) Take x,y in co(A). There exist a_i, b_j in A and $\lambda_i, \mu_j > 0$ ($i = 1, \ldots, m$; $j = 1, \ldots, n$) such that

$$\sum_{i=1}^{m} \lambda_i = \sum_{j=1}^{n} \mu_j = 1,$$

$$x = \lambda_1 a_1 + \ldots + \lambda_m a_m, \quad y = \mu_1 b_1 + \ldots + \mu_n b_n.$$

There is an element c of A such that $c \geqslant a_i$ and $c \geqslant b_j$ for all i,j. Then $c \geqslant x$ and $c \geqslant y$.

(iii) Take x,y in A - P and λ in (0,1). There exist a,b in A such that $x \leq a$ and $y \leq b$. There exists c in A such that $c \geq a$ and $c \geq b$. Then

$$\lambda x + (1-\lambda)y \leq \lambda c + (1-\lambda)c = c,$$

so $\lambda x + (1-\lambda)y \in A - P$.

(iv) Take a,b in A. There exists c in A such that $c \geq a$ and $c \geq b$. Then $f(a) + g(b) \leq (f+g)(c)$.

<u>Cofinal subgroups</u>

Let E be a subgroup of an ordered commutative group. We say that E is <u>cofinal</u> if, given u in P, there exists x in E such that $u \leq x$ (equivalently, if $P \subseteq E - P$). If P generates X, this clearly implies that $E + P = E - P = X$. In general, we have:

<u>1.1.10</u>. If E is a cofinal subgroup, then

$$E + P = E - P = E + P - P,$$

and this set is a subgroup.

<u>Proof</u>. Let $F = E + P - P$. Then F is clearly a subgroup, Take a in F. Then a can be expressed in the form $x + u - v$, where $x \in E$ and $u,v \in P$. Since E is cofinal, there exist y in E and w in P such that $u = y - w$. Hence $a \in E - P$, so $F = E - P$. Thus also $F = -(E - P) = E + P$.

<u>Extremal points of the positive wedge</u>

It is obvious that a wedge has no extreme points except, in the case when it is a cone, zero. For wedges (as opposed to convex sets) the following notion is of more interest: we say that x is an <u>extremal point</u> of a wedge P if each point of the order-interval $[0,x]$ is a positive scalar multiple of x. If $y \in [0,x]$, then $x - y \in [0,x]$, so this implies, in fact, that $y = \lambda x$ for some λ in $[0,1]$. Another way of stating the condition is: pos x is order-convex; pos x is called an <u>extreme ray</u> of P when this is true. If P is a cone, this is clearly equivalent to lin x being order-convex.

<u>Examples</u>. (i) In any of the standard sequence spaces, with the usual ordering, each element e_n is an extremal point of P (where e_n denotes the sequence having 1 in place n and 0 elsewhere). Positive multiples of the e_n are the only extremal points, for if $x = \{\xi_i\}$ is a sequence in P having two terms ξ_i and ξ_j strictly positive, then $0 \leq \xi_i e_i \leq x$, while e_i is not a scalar multiple of x.

(ii) It is easily verified that the usual positive cone in $C[0,1]$ has no non-zero extremal elements.

1.2. Order-convexity

Order convexity will play a basic part in our theory. If $\leq$ is the ordering associated with the semigroup P, then, for any subset A of X,

$$[A] = (A + P) \cap (A - P),$$

so A is order-convex iff $(A + P) \cap (A - P) = A$. (This is a good example of a concept that seems well motivated in its order-theoretic form, but hardly so in its algebraic form.)

We now give some simple properties of order-convex sub-groups and linear subspaces (these are called <u>order-ideals</u> by some writers).

<u>1.2.1.</u> Let E be a subgroup of an ordered commutative group. Then E is order-convex iff $0 \leq x \leq u \in E$ implies $x \in E$.

<u>Proof</u>. The condition is clearly necessary. Suppose that it holds, and that $u \leq x \leq v$, where $u,v \in E$. Then $v - u \in E$ and $0 \leq x - u \leq v - u$, so $x - u \in E$. Hence $x \in E$.

The positive-cone form of the statement in 1.2.1 is as follows: if $x,y \in P$ and $x + y \in E$, then $x,y \in E$.

<u>1.2.2</u>. Let X be a commutative group with an antisymmetric ordering, and let E be a subgroup of X. If $E \cap P = \{0\}$, where P denotes the

positive set, then E is order-convex.

<u>Proof</u>. If $0 \leqslant x \leqslant u \in E$, then $u = 0$, so $x = 0$.

<u>1.2.3</u>. Let A be a subset of an ordered linear space. If A is (i) convex, (ii) a wedge, or (iii) a linear subspace, then so is $[A]$.

<u>Proof</u>. We prove (i), the other cases being similar. Suppose that $a_i, b_i \in A$ and $a_i \leqslant x_i \leqslant b_i$ $(i = 1,1)$. Take λ in $(0,1)$, and write $\lambda' = 1 - \lambda$. Then

$$\lambda a_1 + \lambda' a_2 \leqslant \lambda x_1 + \lambda' x_2 \leqslant \lambda b_1 + \lambda' b_2,$$

so $\lambda x_1 + \lambda' x_2 \in [A]$.

<u>1.2.4</u>. Let E be a subgroup of an ordered commutative group, and define: $x \in J(E)$ iff there exists u in $E \cap P$ such that $-u \leqslant x \leqslant u$. Then $[E] = E + J(E)$.

<u>Proof</u>. Clearly, $J(E) \subseteq [E]$, so $E + J(E) \subseteq [E]$. Take x in $[E]$. There exist u,v in E such that $u \leqslant x \leqslant v$. Then $0 \leqslant x - u \leqslant v - u$, so $x - u \in J(E)$. Hence $x \in E + J(E)$.

<u>1.2.5</u>. If X is an ordered linear space with positive wedge P, and E is a linear subspace of X, then E is order-convex iff $E \cap P$ is an extreme subset of P.

<u>Proof</u>. Straightforward.

<u>Examples</u>

(i) R^2, usual order. If $(\alpha, \beta) > 0$, then the order-interval $[0, (\alpha, \beta)]$ is the rectangle
$\{(\xi, \eta) : 0 \leqslant \xi \leqslant \alpha \text{ and } 0 \leqslant \eta \leqslant \beta\}$.
Among the one-dimensional subspaces, only the axes and the lines with negative gradients are order-convex. The only order-convex, directed proper subspaces are the axes.

(ii) With respect to the usual ordering, the spaces m, c_0, l_1, F are order-convex subsets of s.

(iii) In contrast to the result of 1.2.3, order-convexity is not preserved by algebraic operations. Consider R^3 with the usual order. Let $a = (2, -1, 0)$, $b = (-1, 2, 1)$, $A = \text{lin } a$, $B = \text{lin } b$. Then A and B are order-convex, by 1.2.2, but A + B is not, since $(1,1,1) \in A + B$, while $(1,1,0) \notin A + B$. (It is elementary, however, that $[A] + [B] \subseteq [A + B]$ for any A,B.)
With a,b as above, let $C = \{0,a,b\}$. Then C is order-convex, since no two elements of C are comparable. Now $\frac{1}{2}(1,1,1) = \frac{1}{2}(a + b) \in co(C)$, while $\frac{1}{2}(1,1,0) \notin co(C)$, since this element is not in A + B. Hence co(C) is not order-convex.

1.3. Order-units and Archimedean orderings

Order-units

Let X be an ordered commutative group. An element e of X is said to be an order-unit if, given x in X, there exists a positive integer n such that $-ne \leq x \leq ne$. If X is a linear space, the following condition is equivalent: there exists $\delta > 0$ such that for $0 \leq \lambda \leq \delta$, $-e \leq \lambda x \leq e$, or $e \pm \lambda x \in P$. Hence we have:

1.3.1. Let X be an ordered linear space with positive wedge P. Each of the following statements is equivalent to e being an order-unit in X:

(i) e is an internal point of P:

(ii) $[-e, e]$ is absorbent;

(iii) $[\text{lin } e] = X$.

Hence the set of order-units is the set P_o of internal points of P. With this notation we have:

1.3.2. (i) If $e \in P_o$ and $e \leq f$, then $f \in P_o$;

(ii) If $P \subset X$, then $P_o \cap (-P) = \emptyset$.

<u>Proof</u>. (i) f is an internal point of $P + (f - e)$, which is contained in P.

(ii) If $e \in P_o \cap (-P)$, then $0 = \frac{1}{2}(e - e) \in P_o$, by 0.2.1. Since P is a wedge, this implies that $P = X$.

<u>Examples</u>

(i) m, usual order. Let e be the sequence having every term equal to 1. Then e is an order-unit: if $x = \{\xi_n\}$, and $\sup |\xi_n| = \alpha$, then $-\alpha e \leq x \leq \alpha e$.

(ii) s, usual order. There is no order-unit. For if $x = \{\xi_n\}$, where $\xi_n \geq 0$ for each n, let $\eta_n = n\xi_n + 1$, $y = \{\eta_n\}$. Then, for each n, we have $y \not\leq nx$.

<u>Archimedean orderings and related concepts</u>

Let ω denote the set of positive integers. An ordering of a commutative group X is said to be:

(a) <u>Archimedean</u> if $nx \leq y$ for all n in ω and some y in X implies $x \leq 0$;

(b) <u>almost Archimedean</u> if $-y \leq nx \leq y$ for all n in ω and some y in X implies $x = 0$;

(c) <u>everywhere non-Archimedean</u> if, given x in X, there exists a in P such that $-a \leq nx \leq a$ for all n in ω.

Definitions (b) and (c) were introduced by Bonsall (1).

In linear spaces, there are a number of equivalent forms of the above definitions. The following lemma is useful:

<u>1.3.3. LEMMA</u>. Suppose that x is an element of an ordered linear space X, and that there exist y in X and α, β in R such that $0 < \alpha < \beta$, $x \leq \alpha y$ and $x \leq \beta y$. Then there exists u in P such that $y \leq u$, so that $x \leq \lambda u$ for all $\lambda \geq \alpha$.

<u>Proof</u>. Let $u' = \beta y - x$, $v' = \alpha y - x$. Then $u', v' \in P$ and $(\beta - \alpha)y = u' - v'$. Let $u = (\beta - \alpha)^{-1}u'$. Then $u \geq 0$ and $u \geq y$.

We denote by P_c the lineal closure of P (see Section 0.2).

1.3.4. Let X be an ordered linear space with positive wedge P. Then the following statements are equivalent:

(i) the ordering is Archimedean;

(ii) if $x,y \in X$ and $x \leq \lambda y$ for all $\lambda > 0$, then $x \leq 0$;

(iii) if $x,y \in X$, $\varepsilon > 0$ and $x \leq \lambda y$ for $0 < \lambda < \varepsilon$, then $x \leq 0$;

(iv) if $x,y \in X$, $\varepsilon > 0$ and $x \leq (\alpha + \lambda)y$ for $0 < \lambda < \varepsilon$, then $x \leq \alpha y$;

(v) P is lineally closed.

Proof. (i) implies (ii) a priori. The implications (ii) => (iii) and (iv) => (i) follow immediately from 1.3.3. (iii) => (iv) is clear, on considering $x - \alpha y$. We finish by showing that (v) is equivalent to (iii). (v) implies (iii), since the hypothesis of (iii) implies that $-x \in P_c$. Suppose that (iii) holds, and that $x \in P_c$. Then there exists y in X such that for $0 < \lambda < 1$, $x + \lambda y \in P$, that is, $-x \leq \lambda y$. Hence $x \in P$.

1.3.3 shows that two further equivalent statements are obtained by writing "$y \in P$" instead of "$y \in X$" in (ii) and (iii).

We now have a similar result for almost-Archimedean orderings.

1.3.5. Let X be an ordered linear space with positive wedge P. Then the following statements are equivalent:

(i) the ordering is almost Archimedean;

(ii) if $x,y \in X$ and $-\lambda y \leq x \leq \lambda y$ for all $\lambda > 0$, then $x = 0$;

(iii) if $x,y,z \in X$, $\varepsilon > 0$ and $\lambda y \leq x \leq \lambda z$ for $0 < \lambda < \varepsilon$, then $x = 0$:

(iv) P_c is a cone.

Proof. (i) <=> (ii). If $-y \leq nx \leq y$ for all n in ω, then $y \geq 0$, so $-\lambda y \leq x \leq \lambda y$ for all $\lambda > 0$.

(ii) => (iii). Suppose that $\lambda y \leq x \leq \lambda z$ for $0 < \lambda < \varepsilon$. By 1.3.3, there exist y', z' in P such that $-\lambda y' \leq x \leq \lambda z'$ for all $\lambda > 0$. Let

$u = y' + z'$. Then $-\lambda u \leq x \leq \lambda u$ for all $\lambda > 0$. Hence $x = 0$.

(iii) => (iv). Suppose that $\pm x \in P_c$. Then there exist y,z in X such that, for $0 < \lambda < 1$, $x + \lambda y$ and $-x + \lambda z$ are in P, i.e. $-\lambda y \leq x \leq \lambda z$. Hence $x = 0$.

(iv) => (ii). If $-\lambda y \leq x \leq \lambda y$ for all $\lambda > 0$, then $\pm x + \lambda y \in P$ for all $\lambda > 0$, so $\pm x \in P_c$. Hence $x = 0$.

1.3.6. An ordering of a linear space is almost-Archimedean iff every order-interval is lineally bounded.

Proof. Suppose that all order-intervals are lineally bounded, and that $-y \leq \lambda x \leq y$ for all $\lambda > 0$. Then $x = 0$, since $[-y,y]$ is lineally bounded. Hence the ordering is almost-Archimedean, by 1.3.5 (ii).

Now suppose that some order-interval $[a,b]$ is not lineally bounded, so that there exist $x,y \in X$ such that $x \neq 0$ and $a \leq \lambda x + y \leq b$ for all $\lambda \geq \lambda_o$. Then

$$a - \lambda_o x - y \leq \mu x \leq b - \lambda_o x - y$$

for all $\mu > 0$. Hence, by 1.3.5(iii), the ordering is not almost-Archimedean.

1.3.7. If e is an order-unit, then:

(i) the ordering is Archimedean iff $x \leq \lambda e$ for all $\lambda > 0$ implies $x \leq 0$:

(ii) the ordering is almost-Archimedean iff $-\lambda e \leq x \leq \lambda e$ for all $\lambda > 0$ implies $x = 0$.

Proof. (i) Suppose that $x \leq \lambda y$ for all $\lambda > 0$. There exists $\alpha > 0$ such that $y \leq \alpha e$. Hence $x \leq \lambda e$ for all $\lambda > 0$. (ii) is similar.

Finally, we give two equivalent formulations of the "everywhere non-Archimedean" condition.

1.3.8. Let X be an ordered linear space with positive wedge P. Then the following statements are equivalent:

(i) the ordering is everywhere non-Archimedean;

(ii) given x in X, there exists y in P such that $x \leq \lambda y$ for all $\lambda > 0$;

(iii) $P_c = X$.

<u>Proof</u>. (i) $\Rightarrow$ (iii). Suppose that the ordering is everywhere non-Archimedean, and take x in X. Then there exists a in P such that for all $\lambda > 0$, $-\lambda a \leq x \leq \lambda a$, so that $x + \lambda a \in P$. Hence $x \in P_c$.

(iii) $\Rightarrow$ (ii). Suppose that $P_c = X$, and take x in X. Then there exists z in X such that, for $0 < \lambda < 1$, $-x + \lambda z \in P$, i.e. $x \leq \lambda z$. By 1.3.3, there exists y in P such that $x \leq \lambda y$ for all $\lambda > 0$.

(ii) $\Rightarrow$ (i). Take $x \in X$. There exist y,z in P such that $x \leq \lambda y$ and $-x \leq \lambda z$ for all $\lambda > 0$. Let $a = y + z$. Then $-\lambda a \leq x \leq \lambda a$ for all $\lambda > 0$.

<u>Examples</u>

We shall say that a wedge is Archimedean, etc., if its associated ordering is:

(i) The natural ordering of a function space is Archimedean.

(ii) The lexicographic ordering of R^2 is not almost Archimedean, since $(-1,0) \leq (0,n) \leq (1,0)$ for all n in ω.

(iii) The cone $\{0\} \cup \{(\xi,\eta) : \xi > 0 \text{ and } \eta > 0\}$ in R^2 is almost Archimedean, but not Archimedean.

(iv) The wedge $\{(\xi,\eta) : \xi \geq 0\}$ in R^2 is Archimedean, but not almost Archimedean (clearly, an Archimedean <u>cone</u> is almost Archimedean).

(v) We define an everywhere non-Archimedean ordering of the space F of all finite sequences. Let P be the set of sequences in F whose last non-zero term is positive, together with 0 (the "reverse lexicographic" ordering). Take x in F and suppose that the last non-zero term of x occurs in place k. Let e_{k+1} be the sequence having 1 in place k + 1 and 0 elsewhere. Then $-e_{k+1} \leq nx \leq e_{k+1}$ for all n in ω.

1.4. Direct sums and quotient spaces

Direct sums

If P_i is a wedge in X_i $(i = 1, \ldots, n)$, then $P = P_1 \times \ldots \times P_n$ is a wedge in $X = X_1 \times \ldots \times X_n$. Using the same order notation in each space, the ordering associated with P is given by:

$$(x_1, \ldots, x_n) \leq (y_1, \ldots, y_n) \iff x_i \leq y_i \text{ for each } i.$$

(X,P) is said to be the ordered direct sum of the spaces (X_i, P_i). The following facts are elementary:

1.4.1. P is a cone iff each P_i is a cone. $P - P = X$ iff $P_i - P_i = X_i$ for each i. $(e_1, \ldots, e_n)$ is an order-unit in X iff, for each i, e_i is an order-unit in X_i. P is Archimedean iff each P_i is.

The main problem of interest concerning direct sums is the expression of a given space X as the ordered direct sum of two subspaces E_1, E_2. For this, we require $X = E_1 \oplus E_2$ and $P = (E_1 \cap P) + (E_2 \cap P)$. In our next two results, we find some necessary conditions for this to occur.

1.4.2. Let X be a linear space with an antisymmetric, directed ordering. If E_1, E_2 are linear subspaces whose ordered direct sum is X, then E_1 and E_2 are order-convex and directed.

Proof. Take x in E_1. There exist p,q in P such that $x = p - q$. There exist p_i, q_i in $E_i \cap P$ $(i = 1,2)$ such that $p = p_1 + p_2$, $q = q_1 + q_2$. Then $p_2 - q_2 = (p - q) - (p_1 - q_1) \in E_1$, so $p_2 = q_2$, and $x = p_1 - q_1$. Hence E_1 is directed.

Now suppose that $x, y \in P$ and $x + y \in E_1$. Then there exist x_1, y_1 in $E_1 \cap P$ such that $x = x_1 + x_2$ and $y = y_1 + y_2$. Then $x_2 + y_2 = (x + y) - (x_1 + y_1) \in E_1$, so $x_2 + y_2 = 0$. Hence $x_2 = y_2 = 0$, so $x, y \in E_1$.

<u>1.4.3</u>. Let X be a linear space with an antisymmetric, directed ordering, and let E_1 be an order-convex, directed subspace. Then there exists at most one subspace E_2 such that X is the ordered direct sum of E_1 and E_2.

<u>Proof</u>. E_2 must satisfy $E_2 = (E_2 \cap P) - (E_2 \cap P)$. Further, $x \in E_2 \cap P$ iff $0 \leq y \leq x$ and $y \in E_1$ implies $y = 0$. For if x satisfies this condition, and $x = x_1 + x_2$ $(x_1 \in E_1 \cap P)$, then $x_1 = 0$, so $x \in E_2$.

Later, we shall give conditions which ensure that if X is the direct sum of two order-convex, directed subspaces, then it is the ordered direct sum (section 2.1), and also sufficient conditions for such decompositions to exist (section 2.5).

<u>Quotient spaces</u>

The elementary facts about the ordering induced on a quotient space are summarised in the following theorem:

<u>1.4.4</u>. Let X be an ordered linear space with positive wedge P, and let E be a linear subspace of X. Let r denote the projection onto the quotient space X/E. Then r(P) is a wedge, and the following statements are true concerning the ordering of X/E associated with r(P):

(i) $r(x) \leq r(y)$ iff there exists e in E such that $x \leq y + e$;

(ii) $r(P) - r(P) = X/E$ iff $E + (P - P) = X$;

(iii) $r(x) \in [r(A)]$ iff $x \in [A + E]$;

(iv) r(P) is a cone iff E is order-convex;

(v) r(A) is order-convex iff A + E is order-convex;

(vi) r(a) is an order-unit in X/E iff $[E + \operatorname{lin} a] = X$.

<u>Proof</u>. (i), (ii) Elementary.

(iii) $r(x) \in [r(A)]$ iff there exist a,b in A such that $r(a) \leq r(x) \leq r(b)$. This occurs iff there exist a,b in A and e,f in E such that $a + e \leq x \leq b + f$, that is, iff $x \in [A + E]$.

(iv), (v), (vi) follow.

Archimedean properties of the quotient space bear little relation to those of X, as the following two examples show:

(i) R^2 with its lexicographic ordering is not almost Archimedean. The quotient space with the vertical axis is isomorphic to R with the usual ordering, so is Archimedean.

(ii) Consider m with its usual ordering. F is an order-convex subspace. Let $e = (1,1,\ldots)$, $x = (1,\frac{1}{2}, \ldots, \frac{1}{n}, \ldots)$. For each n, there exists e_n in F such that $-e \leq nx \leq e + e_n$, so $-r(e) \leq nr(x) \leq r(e)$. Since $x \notin F$, this shows that m/F is not almost Archimedean.

1.5. Linear mappings and functionals

Let X,Y be ordered linear spaces, with positive wedges P,Q respectively. It is clear that a linear mapping f from X to Y is monotonic iff $f(P) \subseteq Q$. We denote the set of monotonic linear mappings by (P,Q). This is a wedge, inducing a linear ordering on the space L(X,Y) of all linear mappings from X to Y. In order notation,

$$f \leq g \Leftrightarrow f(x) \leq g(x) \text{ for all } x \text{ in } P.$$

In particular, when Y = R, we obtain the ordering of X' associated with P^o. This will be called the "dual ordering" of X'.

Monotonic linear mappings and functionals are called "positive" by many writers. The next few results summarise some of their elementary properties.

1.5.1. If P - P = X and Q is a cone, then (P,Q) is a cone.

Proof. Obvious.

Conversely, it is easily seen that if (P,Q) is a cone, then P - P = X and Q is a cone.

1.5.2. Suppose that Q is Archimedean. Then:

(i) (P,Q) is Archimedean;

(ii) if $f \in L(X,Y)$ and $f(P)$ is bounded below, then $f \in (P,Q)$.

Proof. (i) Suppose that $f,g \in L(X,Y)$ and $nf \leq g$ $(n \in \omega)$. For x in P, $nf(x) \leq g(x)$ $(n \in \omega)$, so $f(x) \leq 0$. Hence $f \leq 0$.

(ii) Suppose that $f(P) \geq y$. Then, for x in P, $nf(x) \geq y$ $(n \in \omega)$, so $f(x) \geq 0$.

1.5.3. If f is a monotonic linear mapping on X onto Y, and e is an order-unit in X, then $f(e)$ is an order-unit in Y.

Proof Given y in Y, there exists x in X such that $f(x) = y$. For some n in ω, $-ne \leq x \leq ne$. Then $-nf(e) \leq y \leq nf(e)$.

Even if f does not map onto Y, 1.5.3 shows that, provided that $f(P) = -f(P)$, we have $f(e) \not\leq 0$.

1.5.4. If $-x \leq y \leq x$ and $-f \leq g \leq f$, then $g(y) \leq f(x)$.

Proof. We have $f(x+y) \geq g(x+y)$, or $g(y) \leq f(x) + f(y) - g(x)$. Also, $g(x-y) \geq -f(x-y)$, or $g(y) \leq g(x) + f(x) - f(y)$. Adding the two inequalities, we obtain $g(y) \leq f(x)$.

It is clear that the inverse image of an order-convex set under a monotonic mapping is order-convex. In particular, if Q is a cone and $f \in (P,Q)$, then the kernel of f is order-convex. For linear functionals, the converse holds:

1.5.5. Suppose that f is a linear functional with an order-convex kernel. Then either f or $-f$ is monotonic.

Proof. If neither f nor $-f$ is monotonic, then there exist x,y in P such that $f(x) = 1$, $f(y) = -1$. Then $x + y$ is in the kernel of f, while x is not. Hence the kernel of f is not order-convex.

Next, we give an elementary extension theorem which will be used repeatedly.

1.5.6. Suppose that X,Y are linear spaces, and that P is a wedge in X. If f is an additive, positive homogeneous mapping from P to Y, then f can be extended to a linear mapping from X to Y.

Proof. It is sufficient to show that f has a linear extension to P - P. Suppose that u - v = u' - v', where u,v,u',v' ∈ P. Then u + v' = u' + v, so f(u) + f(v') = f(u') + f(v), or f(u) - f(v) = f(u') - f(v'). Hence we can extend f to P - P by defining: f(u - v) = f(u) - f(v). The mapping so defined is clearly additive and positive homogeneous. It also satisfies f(-x) = -f(x), so is linear.

Order-bounded linear mappings

If X,Y are ordered linear spaces, a linear mapping from X to Y is said to be order-bounded if it maps order-intervals into order-intervals. Clearly, it is sufficient if it maps each order-interval of the form [0,a] into an order-interval. Monotonic linear mappings are order-bounded. From the next result, we see that the same is true of linear mappings of the form f - g, where f and g are monotonic.

1.5.7. If X,Y are ordered linear spaces, then the order-bounded linear mappings form a linear subspace of L(X,Y).

Proof. It is obvious that if f is order-bounded and $\lambda \in R$, then λf is order-bounded (though positive and negative scalars must be considered separately). Suppose that f,g are order-bounded, and take an order-interval $[x_1,x_2]$ in X. Then there exist y_i,z_i in Y (i = 1,2) such that $f[x_1,x_2] \subseteq [y_1,y_2]$ and $g[x_1,x_2] \subseteq [z_1,z_2]$ By 1.1.5, $(f + g)[x_1,x_2] \subseteq [y_1 + z_1, y_2 + z_2]$.

An elegant result concerning order-bounded linear functionals is:

1.5.8. Suppose that f is an order-bounded linear functional on X. Then, for x > 0,

$$\sup f[0,x] + \inf f[0,x] = f(x).$$

Proof. Let $\sup f\,[0,x] = \alpha$, $\inf f\,[0,x] = \beta$. If $y \in [0,x]$, then $x - y \in [0,x]$, so

$$\beta \leq f(x-y) \leq \alpha,$$

or

$$f(x) - \alpha \leq f(y) \leq f(x) - \beta.$$

The right-hand inequality shows that $\alpha \leq f(x) - \beta$, while the left-hand inequality shows that $\beta \geq f(x) - \alpha$. Hence $\alpha + \beta = f(x)$.

Examples

(i) R^n, usual ordering. Corresponding to (α_i) in R^n is the linear functional f defined by $f(x) = \Sigma \alpha_i \xi_i$, where $x = (\xi_i)$. Clearly, f is monotonic iff each $\alpha_i \geq 0$.

(ii) R^2, lexicographic ordering. Let $f(\xi,\eta) = \alpha\xi + \beta\eta$. If $e_1 = (1,0)$, the order-interval $[-e_1, e_1]$ contains $(0,\lambda)$ for all λ in R. Hence we see that f is order-bounded iff $\beta = 0$.

(iii) The inverse of a one-to-one monotonic linear mapping need not be monotonic. Let X be R^2 with the usual ordering, and let $f(\xi,\eta) = (\xi, \xi + \eta)$. Then f^{-1} is not monotonic, since $f^{-1}(1,0) = (1,-1)$.

(iv) If X is everywhere non-Archimedean, then there are clearly no non-zero order-bounded linear functionals on X.

(v) We give an example where the ordering is antisymmetric and Archimedean, but where there are no non-zero monotonic linear functionals.

Take p such that $0 < p < 1$, and let X be the space $L_p\,[0,1]$ (with respect to Lebesgue measure μ), with the natural ordering. Write $q(x) = \int_0^1 |x|^p$. Suppose that there is a non-zero monotonic linear functional f, so that there exists a non-negative function x_0 such that $q(x_0) = 1$ and $f(x_0) = \alpha > 0$. Since $\int_0^s x_0^p$ is a continuous function of s, there exist disjoint intervals I_1, I_1', with union $[0,1]$, such that

$$q(x_o \chi_1) = q(x_o \chi_1') = \tfrac{1}{2},$$

$$f(x_o \chi_1) \geq \frac{\alpha}{2},$$

where χ_1, χ_1' denote the characteristic functions of I_1, I_1'. We can arrange also that I_1' contains $\{s : x_o(s) = 0\}$. Define $x_1 = x_o + x_o \chi_1$.

Repeating this process, we can construct sequences of intervals $\{I_n\}$, $\{I_n'\}$ and functions x_n such that:

$$I_n \cup I_n' = I_{n-1}, \quad I_n \cap I_n' = \emptyset,$$

$$q(x_{n-1} \chi_n) = q(x_{n-1} \chi_n') = \frac{1}{2} c^{n-1},$$

$$f(x_{n-1} \chi_n) \geq \frac{\alpha}{2},$$

$$x_n = x_{n-1} + x_{n-1} \chi_n,$$

where $c = 2^{p-1}$ and χ_n, χ_n' are the characteristic functions of I_n, I_n'. We notice that x_n is obtained from x_{n-1} by doubling its value on I_n, so that $q(x_n \chi_n) = 2^p q(x_{n-1} \chi_n) = c^n$.

For each n,

$$x_n = x_o \chi_1' + x_1 \chi_2' + \ldots + x_{n-1} \chi_n' + x_n \chi_n,$$

so

$$q(x_n) = \tfrac{1}{2}(1 + c + \ldots + c^{n-1}) + c^n$$

$$\to \quad \frac{1}{2(1 - c)} \quad \text{as } n \to \infty$$

Let $I = \bigcap_{n=1}^{\infty} I_n$. Clearly, $\{x_n(s)\}$ is convergent for s not in I. If $\mu(I) > 0$, then there exists $\varepsilon > 0$ such that $\mu\{s \in I : x_o(s) > \varepsilon\} > 0$. But $x_n(s) = 2^n x_o(s)$ for s in I, so this contradicts the fact that $q(x_n)$ is bounded. Hence the increasing sequence $\{x_n\}$ is convergent almost everywhere to an element x of $L_p[0,1]$, and $f(x) \geq f(x_n)$ for all n. But this is impossible, since $f(x_n) - f(x_{n-1}) \geq \frac{\alpha}{2}$ for each n.

Alternatively, we can argue as follows. X is a complete metric space with respect to the metric $d(x,y) = q(x - y)$, and it follows from a later theorem (3.5.5) that any monotonic linear functional on X is continuous with respect to d. But it is known (see, e.g, Köthe (1), p. 162) that there are no non-zero continuous linear functionals on X (but the proof of this last statement is basically similar to the argument given above).

Half-spaces

By a half-space we mean a set of the form $\{x : f(x) \geq 0\}$, where f is a non-zero linear functional. If P denotes this set, it is clear that P is a wedge, that $P \cap (-P)$ is a maximal subspace, and that the set P_o of internal points of P is $\{x : f(x) > 0\}$, so that $P \cup (-P_o) = X$. Conversely, we have:

1.5.9 THEOREM. If P is a proper wedge such that $P \cup (-P_o) = X$, then P is a half-space.

Proof. The hypothesis implies that $P_o \neq \emptyset$. Recall that, by 1.3.2(ii), $P_o \cap (-P) = \emptyset$. Let $F = P \cap (-P)$. Take e in P_o, x in X. There exists α in R such that $x - \alpha e \in P$. If $\beta < \alpha$, then $x - \beta e = (x - \alpha e) + (\alpha - \beta)e \in P_o$, so $\beta e - x \notin P$. Hence we can define

$$p(x) = \inf \{\lambda \in R : x \leq \lambda e\}.$$

Take x in X, and let $y = x - p(x)e$. We show that $\pm y \notin P_o$, from which it follows that $y \in F$. If $y \in P_o$, then there exists $\varepsilon > 0$ such that $\varepsilon e \leq y$, or $(p(x) + \varepsilon)e \leq x$. But $(p(x) + \varepsilon)e - x \in P_o$, so this is impossible. If $-y \in P_o$, then there exists $\varepsilon > 0$ such that $\varepsilon e \leq -y$, or $x \leq (p(x) - \varepsilon)e$, which contradicts the definition of p. Hence $y \in F$, as stated. Therefore p is linear, and $p(x) \geq 0$ implies $x \in P$. If $p(x) < 0$, then $x \in -P_o$, so $x \notin P$. This completes the proof.

1.5.10 COROLLARY. If P is a proper, Archimedean wedge such that $P \cup (-P) = X$, then P is a half-space.

Proof. P is lineally closed, so if $x \notin P$, then x is an internal point of $X - P$, and hence of $-P$.

1.5.11. COROLLARY. If P is a cone, and

either (i) $P \cup (-P_0) = X$,

or (ii) P is almost Archimedean and $P \cup (-P) = X$,

then X is one-dimensional.

Proof. In both cases, we show that P is a half-space, so that $\{0\}$ is a maximal subspace. In case (i), this follows from 1.5.9. Case (ii) follows from 1.5.10 if we show that P is Archimedean. Suppose that $nx \leq y$ $(n \in \omega)$. Either $x \leq 0$ or $x \geq 0$. If $x \geq 0$, then $-y \leq nx \leq y$ $(n \in \omega)$, so $x = 0$.

1.6. Extension and separation theorems

(1) Using the Hahn-Banach theorem

The basic theorem on extension of monotonic linear functionals is the following:

1.6.1. THEOREM (Bauer, Bonsall, Namioka). Suppose that E is a cofinal linear subspace of an ordered linear space X. Then a monotonic linear functional defined on E has a monotonic linear extension to X.

Proof. Let $X_1 = E + P - P$. It is sufficient to show that f has a monotonic extension to X_1, for then we can take any extension to X. Take $x \in X_1$. By 1.1.10, there exist z, z' in E such that $z \leq x \leq z'$. Hence we can define

$$p(x) = \inf \{f(y) : y \in E \text{ and } y \geq x\},$$

and $p(x) \geq f(z)$. It is easily checked that p is sublinear. For $y \in E$, $p(y) = f(y)$. By the Hahn-Banach theorem, f has an extension $\overline{f}$ defined on X_1 and dominated by p there. If $x \leq 0$, then $p(x) \leq 0$, so $\overline{f}(x) \leq 0$. Hence $\overline{f}$ is monotonic.

If e is an order-unit, then a linear subspace containing e is cofinal, so we have:

1.6.2. COROLLARY (Krein-Rutman). Let X be an ordered linear space, and E a linear subspace containing an order-unit. Then a monotonic linear functional defined on E has a monotonic extension to X.

If $P \subset X$ and e is an order-unit, then $-e \notin P$ (1.3.2), so $f(\lambda e) = \lambda$ defines a monotonic linear functional on lin e, and we have:

1.6.3. COROLLARY. If P is a proper wedge with internal points, then $P^{\circ} \neq \emptyset$.

A simple example shows that monotonic linear functionals defined on non-cofinal subspaces do not necessarily have monotonic extensions. Taking R^2 with the lexicographic ordering, a monotonic linear functional on the vertical axis is defined by: $f(0, \eta) = \eta$. No extension of this functional is monotonic (cf. ex. (ii), section 1.5).

The following result, due to Bauer, gives a theoretical answer to the question of when monotonic extensions exist.

1.6.4. Suppose that E is a linear subspace of an ordered linear space X, and that f is a monotonic linear functional defined on E. Then the following statements are equivalent:

(i) f has a monotonic linear extension to X;

(ii) there exists a convex, absorbent set U such that $f \leq 1$ on $E \cap (U - P)$.

Proof. (i) => (ii). Let $U = \{x : \overline{f}(x) \leq 1\}$, where $\overline{f}$ is a monotonic extension of f.

(ii) => (i) Let p be the Minkowski functional of $U - P$. For $y \in E$, $f(y) \leq p(y)$. By the Hahn-Banach theorem, f has an extension $\overline{f}$ defined on X and dominated by p there. If $x \leq 0$, then $p(x) = 0$, so $\overline{f}(x) \leq 0$. Hence $\overline{f} \in P^{\circ}$.

Assuming Eidelheit's separation theorem (0.2.4), it is easy to deduce the following theorem on separation by monotonic linear functionals:

1.6.5. Let X be a real linear space, P a wedge in X, and A a convex subset of X. If $x \in X$, then the following statements are equivalent:

(i) there exists f in P^o such that $f(x) > \sup f(A)$;

(ii) there exists a convex, absorbent set U such that $(x + U) \cap (A - P) = \emptyset$.

<u>Proof</u>. (i) $\Rightarrow$ (ii). Let $U = \{x : |f(x)| < \delta\}$, where $\delta = f(x) - \sup f(A)$.

(ii) $\Rightarrow$ (i). The hypothesis of (ii) implies that there is a non-zero linear functional f on X such that $f(x + U) \geq f(A-P)$. Then $f(P)$ is bounded below, so $f \in P^o$. Since U is absorbent, there exists u in U such that $f(u) < 0$. It follows that $f(x) > \sup f(A)$.

<u>(2) Direct methods</u>

Alternatively, we can give a direct proof of the monotone extension theorem 1.6.1, and deduce the standard extension and separation theorems.

<u>1.6.1. Alternative proof</u> (Day). Suppose that E is a cofinal linear subspace, that f is a monotonic linear functional defined on E, and that $x \in X \sim E$. We show that there is a monotonic extension of f to $E + \text{lin } x$. An application of Zorn's lemma then gives the result.

If $x \notin E + P - P$, $y \in E$ and $y + \lambda x \in P$, then $\lambda = 0$, so any extension of f to $E + \text{lin } x$ is monotonic.

Suppose that $x \in E + P - P$. Let $A = E \cap (x - P)$, $B = E \cap (x + P)$. By 1.1.10, A and B are non-empty. Let $\alpha = \sup f(A)$, $\beta = \inf f(B)$. Then $\alpha \leq \beta$ Take γ in$[\alpha, \beta]$, and define

$$f(y + \lambda x) = f(y) + \lambda\gamma \qquad (y \in E).$$

Suppose that $y \in E$ and $y - \lambda x \in P$. We must show that $f(y) \geq \lambda\gamma$. If $\lambda > 0$, then $\lambda^{-1}y \in B$, so $f(y) \geq \lambda\beta \geq \lambda\gamma$. If $\lambda < 0$, then $x - \lambda^{-1}y \in P$, so $\lambda^{-1}y \in A$, and $f(\lambda^{-1}y) \leq \alpha$, or $f(y) \geq \lambda\alpha \geq \lambda\gamma$.

<u>Deduction of the Hahn-Banach theorem</u> (0.2.2). Let p be a sublinear functional on X, and let f be a linear functional defined

on a subspace E of X and dominated by p there. Define subsets of $X \times R$ as follows:

$$E_1 = E \times R,$$
$$P = \{(x,\lambda) : \lambda \geq p(x)\}.$$

Then P is a wedge, and $E_1 + P = X \times R$, since for any x in X and λ in R, $(x,p(x)) \in P$ and $(0,\lambda - p(x)) \in E_1$. For x in E, let $g(x,\lambda) = \lambda - f(x)$. Then g is defined on E_1, and $g \geq 0$ on $E_1 \cap \dot{P}$. By 1.6.1, g has an extension $\bar{g}$ defined on $X \times R$ and taking non-negative values on P. Let $\bar{f}(x) = -\bar{g}(x,0)$ $(x \in X)$. Then $\bar{f}(x) = f(x)$ $(x \in E)$, and

$$p(x) - \bar{f}(x) = \bar{g}(x,p(x)) \geq 0 \quad (x \in X).$$

<u>Deduction of Eidelheit's separation theorem</u> (0.2.4) from 1.6.3. Suppose that A,B are convex sets such that $A_o \neq \emptyset$ and $A_o \cap B = \emptyset$. Let $P = pos(A_o - B)$. Then P is a wedge with internal points. We show that P is a proper subset of X; 1.6.3 then says that P^o contains a non-zero element f, and for this f we will have $\inf f(A) = \inf f(A_o) \geq \sup f(B)$. Take a in A_o and b in B. We show that $b - a \notin P$. Now if $b - a \in P$, there exist a' in A_o, b' in B and $\lambda > 0$ such that $b - a = \lambda(a' - b')$, or $a + \lambda a' = b + \lambda b'$. But then, writing $(1 + \lambda)^{-1} = \mu$, we have $\mu(a + \lambda a') \in A_o$ (by 0.2.1) and $\mu(b + \lambda b') \in B$, contradicting the hypothesis that A_o is disjoint from B.

In this section, we give two variants (1.7.1, 1.7.3) of what may be regarded as the central theorem of the whole subject. The proof given for 1.7.1 is due to Weston (1), while that given for 1.7.3 is an extension of a method due to Namioka. Either method can be adapted, in fact, to give proofs of both results. Other proofs of 1.7.1, its corollary 1.7.2, or related results, have been given by Bonsall, Schaefer, Bauer and Riedl. The full strength of 1.7.3 was first proved by Grosberg and Krein (1).

<u>1.7.1. THEOREM</u>. Let X be an ordered linear space with positive wedge P, and let U be a convex, absorbent subset of X. Suppose that f is a linear functional on X such that, for some $\alpha > 0$, $\sup f[-x,x] \leq \alpha$ for all x in $P \cap U$. Then there exists g in X' such that $-g \leq f \leq g$ and $\sup g(U) \leq \alpha$.

<u>Proof</u>. Since U is absorbent, we can define

$$q(x) = \sup f[-x, x\,] \ (x \in P),$$

and $q(x) \leq \alpha$ for x in $P \cap U$. It is easily verified that q is superlinear on P. If $f = 0$ on $P \cap U$, then $f = 0$ on P, and the result holds with $g = 0$. We suppose that this is not the case. Let

$$Q = \{x \in P : q(x) > \alpha\,\}.$$

Then Q is non-empty and convex, and $Q \cap U = \emptyset$. Therefore there exists g in X' such that $g \leq \alpha$ on U and $g \geq \alpha$ on Q. Take x in P. If $q(x) > 0$, it is clear that $g(x) \geq q(x)$. If $q(x) = 0$, take y in Q. Then $nx + y \in Q$ for all n in ω, so $ng(x) + g(y) \geq \alpha$ $(n \in \omega)$, and $g(x) \geq 0$. In both cases, we have $g(x) \geq q(x) \geq \pm f(x)$.

By a similar argument we can show that if $\sup f[0,x] \leq \alpha$ for all x in $P \cap U$, then there exists g in X' such that $g \geq f$, $g \geq 0$ and $\sup g(U) \leq \alpha$.

<u>1.7.2. COROLLARY</u>. Let X be an ordered linear space with positive wedge P, and let f be a linear functional on X. Then the following statements are equivalent:

(i) $f \in P^o - P^o$.

(ii) There exists a convex, absorbent set U such that f is bounded above on $P \cap (U - P)$.

<u>Proof</u>. (i) => (ii). Suppose that $f = g - h$, where $g,h \in P^o$. Let

$$U = \{x \in X : |g(x)| \leq 1 \text{ and } |h(x)| \leq 1\}.$$

Then U is convex and absorbent. If $y \in P \cap (U - P)$, then $0 \leq y \leq x$ for some x in U, so $f(y) \leq g(y) \leq g(x) \leq 1$.

(ii) => (i). Suppose that $f \leq \alpha$ on $P \cap (U - P)$ (where $\alpha > 0$), and let $V = \{x \in U : |f(x)| \leq \alpha\}$. Take x in $P \cap V$. If $-x \leq y \leq x$, then $\frac{1}{2}(x + y) \in P \cap (V - P)$, so $f(x + y) \leq 2\alpha$, and $f(y) \leq 3\alpha$. By 1.7.1, there exists g in X' such that $-g \leq f \leq g$. Then g and $g - f$ are in P^o, so $f \in P^o - P^o$.

1.7.3. THEOREM. Let X be an ordered linear space with positive wedge P, and let U be a convex, symmetric, absorbent, order-convex subset of X. Suppose that f is a linear functional on X such that $f(U)$ is bounded. Then there exist g,h in P^o such that $f = g - h$ and

$$\sup g(U) + \sup h(U) = \sup f(U).$$

Proof. Let $\alpha = \sup f(U)$. We suppose that f is non-zero, so that $\alpha > 0$. Let $\hat{X} = X \times X \times R$, and define subsets of $\hat{X}$ as follows:

$$Q = P \times P \times R^+,$$

$$E = \{ (x, -x, -f(x)) : x \in X\},$$

$$V = \{ (u,v,\lambda) : u,v \in U \text{ and } |\lambda| < \alpha \}.$$

Then Q is a wedge, E is a linear subspace, and V is convex, symmetric and absorbent. If $x + u \in P$ and $-x + v \in P$, then $-u \leq x \leq v$, so $x \in U$ and $|f(x)| \leq \alpha$. Hence if

$$(x, -x, -f(x)) + (u,v,\lambda) \in Q$$

for some u,v in U, then $\lambda + \alpha \geq 0$. It follows that $E + V - (0,0,2\alpha)$ is disjoint from Q, so that there is a non-zero linear functional φ on $\hat{X}$ such that $\varphi \geq 0$ on Q and $\varphi \leq 0$ on $E + V - (0,0,2\alpha)$. Since φ is non-zero, there are points of V at which it takes positive values, so $\varphi(0,0,2\alpha) > 0$. Multiplying by a suitable positive scalar, we may assume that $\varphi(0,0,1) = 1$. Then $\sup \varphi(V) \leq 2\alpha$, and since $\varphi(E)$ is bounded above, $\varphi(E) = \{0\}$. Let

$$\varphi(x, 0, 0) = g(x), \quad \varphi(0, x, 0) = h(x) \quad (x \in X).$$

Then $\varphi(x, y,\lambda) = g(x) + h(y) + \lambda$ for all (x, y,λ) in $\hat{X}$, and it is immediate that g and h are in P^o. Since $\varphi = 0$ on E, we have

$$g(x) - h(x) = f(x) \qquad (x \in X).$$

If $u,v \in U$ and $0 < \varepsilon < \alpha$, then $(u, v,\alpha - \varepsilon) \in V$, so

$$p(u, v, \alpha-\varepsilon) = g(u) + h(v) + \alpha - \varepsilon \leq 2\alpha,$$

or $g(u) + h(v) \leq \alpha + \varepsilon$. Hence $\sup g(U) + \sup h(U) \leq \alpha$. The reverse inequality is obvious, and this completes the proof.

Although, in all of our "natural" examples, P generates X, it is by no means always the case that P^o generates X'. This will become more evident in Chapter 3 (we shall see that, in most cases, $P^o - P^o$ coincides with the space of linear functionals that are continuous with respect to the "usual" topology for the space considered).

One case in which P^o does generate X' is when X is the space F of all finite real sequences, with the usual ordering. Then X' is isomorphic in the obvious way to the space s of all real sequences, and P^o corresponds to the usual positive cone in s.

<u>An order-bounded linear functional not in $P^o - P^o$</u>

The following example is due to Namioka (1). It is slightly awkward, but its inclusion is justified by its theoretical importance.

Let X be $L_p[0,1]$, where $0 < p < 1$, and let Q be the usual positive cone in X. We have seen (section 1.5, ex. (v)), that $Q^o = \{0\}$. Let e be the function defined by $e(s) = 1$ $(0 \leq s \leq 1)$. Then $-e$ is an internal point of $X \sim Q$. On the other hand, since $Q^o = \{0\}$, we know that $U - e$ meets Q for each convex, absorbent set U. Let $P = \text{pos}(Q + e)$. Since 0 is an internal point of $X \sim (Q + e)$, we can define

$$q(x) = \sup\{\lambda > 0 : \lambda^{-1}x \in Q + e\} \qquad (x \in P).$$

It is easily checked that q is superlinear, so that

$$P_1 = \{(x,\lambda) : x \in P \text{ and } q(x) + \lambda \geq 0\}$$

is a wedge in $X \times R$. Let $X \times R$ have the corresponding ordering, and let f be the linear functional defined on $X \times R$ by $f(x,\lambda) = \lambda$. If $0 \leq (y,\mu) \leq (x,\lambda)$, then $(\lambda - \mu) + q(x - y) \geq 0$, so $\mu \leq \lambda + q(x)$, since $q(x) \geq q(x - y) + q(y)$. Also, $\mu \geq -q(y) \geq -q(x)$. Hence

$$-q(x) \leq f(y,\mu) \leq \lambda + q(x),$$

and f is order-bounded.

Given a convex, absorbent set U_1 in $X \times R$, let $U = \{x : (x,0) \in U_1\}$ Then U is a convex, absorbent subset of X, so for each n in ω, there is a point x_n in $U \cap n(Q + e)$. Then $q(x_n) \geq n$, so $0 \leq (0,n) \leq (x_n,0)$. Hence f is unbounded on $P_1 \cap (U_1 - P_1)$, so $f \notin P_1^o - P_1^o$, by 1.7.2.

1.8. Extremal monotonic linear functionals

Characterisation

1.8.1. THEOREM (Hayes). Let X be an ordered linear space with a generating positive wedge P. If f is a non-zero element of P^o, then the following statements are equivalent:

(i) f is an extremal element of P^o.

(ii) Given $\varepsilon > 0$ and x such that $f(x) = 0$, there exists a such that $a \geq 0$, $a \geq x$ and $f(a) \leq \varepsilon$.

(iii) For all x,y in X,

$$\inf \{f(z) : z \geq x \text{ and } z \geq y\} = \max (f(x), f(y)).$$

Proof. (i) => (ii). Since P generates X, we can define

$$q(x) = \inf \{f(y) : y \geq 0 \text{ and } y \geq x\} \qquad (x \in X),$$

and q is sublinear. Take x_0 such that $f(x_0) = 0$. By the Hahn-Banach theorem, there exists g in X' such that $g(x) \leq q(x)$ $(x \in X)$ and $g(x_0) = q(x_0)$. For $x \geq 0$, $g(x) \leq q(x) = f(x)$, while for $x \leq 0$, $g(x) \leq q(x) = 0$. Hence $0 \leq g \leq f$, so g is a scalar multiple of f. Thus $g(x_0) = 0$, so $q(x_0) = 0$, which implies statement (ii).

(ii) => (iii). Take x,y in X and $\varepsilon > 0$. Suppose that $f(x) - f(y) = \lambda \geq 0$. There exists c in P such that $f(c) = 1$. Then $f(x - y - \lambda c) = 0$, so, by (ii), there exists a in P such that $a \geq x - y - \lambda c$ and $f(a) \leq \varepsilon$. Let $z = a + y + \lambda c$. Then $z \geq x$, $z \geq y$ and $f(z) \leq f(x) + \varepsilon$.

(iii) => (i). Suppose that $0 \leq g \leq f$. We show that if $f(x) = 0$, then $g(x) = 0$, which will imply statement (i). Take $\varepsilon > 0$ and x such that $f(x) = 0$. By (iii), there exists a in P such that $a \geq x$ and $f(a) \leq \varepsilon$. Then $g(x) \leq g(a) \leq \varepsilon$. This is true for all $\varepsilon > 0$, so $g(x) \leq 0$. Hence also $g(-x) \leq 0$, so $g(x) = 0$.

Existence

Just as the decomposition of linear functionals into monotonic components depends on order-convex, absorbent sets, the existence of extremal monotonic functionals depends on upward-directed, absorbent sets. The following existence theorem is essentially a generalisation of the main step leading to Kakutani's representation of M-spaces (cf. section 4.3). We prove it with the aid of 0.3.6. If this result is regarded as a corollary of the Krein-Mil'man theorem, then we must admit to resorting to topological methods in this purely algebraic chapter. It is therefore of interest to record that an algebraic proof of 0.3.6 (using, in fact, the methods of section 1.10) has been given by Bonsall (3).

1.8.2. THEOREM. Let X be an ordered linear space with positive wedge P, and suppose that K is an upward-directed, absorbent subset of X such that $K - P = K$. Let p be the Minkowski functional of K. Then, for each x_o in X, there is an extremal element f of P^o such that $f(x) \leq p(x)$ $(x \in X)$ and $f(x_o) = p(x_o)$.

Proof. By 1.1.9 (iii), K is convex, so p is sublinear. A linear functional f is dominated by p iff $f \leq 1$ on K: denote the set of such functionals by H. Since $K - P = K$, each element of H is bounded below on P, and is therefore in P^o. We show that a non-zero extreme point of H is an extremal point of P^o. The result then follows, by 0.3.6.

Let f be a non-zero extreme point of H, and write $\sup f(K) = \lambda$. Then $\lambda > 0$, since f is non-zero and K is absorbent. If $\lambda < 1$, then $f/\lambda \in H$, contradicting the fact that f is an extreme point. Hence $\lambda = 1$. Suppose that $f = g + h$, where g,h are non-zero elements of P^o. Let $\sup g(K) = \mu$. Then $\sup h(K) = 1 - \mu$, by 1.1.9 (iv), and $0 < \mu < 1$, since g and h are non-zero. Let $g' = g/\mu$, $h' = h/(1 - \mu)$. Then

$g', h' \in H$ and $f = \mu g' + (1 - \mu)h'$. Hence $g' = f$, or $g = \mu f$.

<u>1.8.3. COROLLARY</u>. Let X be an ordered linear space with a positive wedge P that is a proper subset of X and contains an order-unit e. Let $p(x) = \inf\{\lambda > 0 : x \leq \lambda e\}(x \in X)$. Then for each x_o in X, there is an extremal element f of P^o such that $f(x) \leq p(x) \quad (x \in X)$ and $f(x_o) = p(x_o)$. If the ordering is almost Archimedean, then the extremal elements of P^o separate points of X.

<u>Proof</u>. Let $K = e - P$. Then K is upward-directed and absorbent, and $K - P = K$. Also, p is the Minkowski functional of K. The first statement follows.

Suppose that the ordering is almost Archimedean. If $p(x) = p(-x) = 0$, then $-\lambda e \leq x \leq \lambda e$ for all $\lambda > 0$, so $x = 0$. Thus if $x \neq 0$, then either $p(x)$ or $p(-x)$ is non-zero, so there is an extremal element of P^o which is non-zero at x.

<u>Extension</u>

<u>1.8.4</u>. Suppose that X is a directed ordered linear space, and that E is a cofinal subspace. If f is an extremal monotonic linear functional on E, then there is an extremal monotonic linear functional on X that extends f.

<u>Proof</u>. Modify the second proof of 1.6.1 by taking $\gamma = \beta$ instead of $\gamma \in [\alpha,\beta]$. We show that this gives an extremal monotonic linear functional on E + lin x. An application of Zorn's lemma to the set of extremal monotonic extensions of f then gives the result. Suppose that g is a linear functional on E + lin x such that $0 < g \leq f$. Then there exists μ in $(0,1]$ such that $g(y) = \mu f(y) \quad (y \in E)$. The proof is completed by showing that $g(x) = \mu\beta$. For all y in B, we have

$$0 \leq g(y - x) = \mu f(y) - g(x),$$

so $g(x) \leq \mu\beta$. Also,

$$0 \leq f(y - x) - g(y - x) = (1 - \mu)f(y) - \beta + g(x),$$

so $\beta - g(x) \leq (1 - \mu)\beta$, or $g(x) \geq \mu\beta$.

1.9. Bases for cones

Definition. A *base* for a wedge P is a convex subset B such that, for each x in P ~ {0}, there exists a unique positive number f(x) such that $x/f(x) \in B$.

It follows from convexity of B and uniqueness of f(x) that $0 \notin B$, and hence, by 1.1.6, that P must be a cone.

A linear functional f is said to be *strictly monotonic* if $f(x) > 0$ whenever $x > 0$. We show that a cone P has a base iff there is a strictly monotonic linear functional, the correspondence being one-to-one if P generates X. If f is strictly monotonic, it is clear that $\{x \in P : f(x) = 1\}$ is a base for P. The converse is given by the following result:

1.9.1. Let B be a base for the cone P, and, for x in P~ {0}, let f(x) be the number such that $x/f(x) \in B$. Then f can be extended to a strictly monotonic linear functional on X.

Proof. Clearly, $f(\lambda x) = \lambda f(x)$ for $x \in P$, $\lambda > 0$. Take x,y in P ~ {0}, and let $f(x) + f(y) = \alpha$. Then

$$\alpha^{-1}(x + y) = \alpha^{-1}f(x)(x/f(x)) + \alpha^{-1}f(y)(y/f(y)) \in B,$$

so $f(x + y) = \alpha$. By 1.5.6, f has a linear extension to X.

Examples.

(i) l_1, usual order. A strictly monotonic linear functional is defined by $f(x) = \sum_1^\infty \xi_n$.

(ii) m, usual order. A strictly monotonic linear functional is defined by $f(x) = \sum_1^\infty 2^{-n} \xi_n$.

(iii) If e is an order-unit in X, then $\{f \in P^\circ : f(e) = 1\}$ is a base for P°, by 1.5.3.

(iv) s, usual order. There is no strictly monotonic linear functional (and hence no base). For if $f(e_n) = \alpha_n > 0$ for each n, and $a = \{ n\alpha_n^{-1} \}$, then monotonicity of f gives $f(a) \geq n$ for each n. (This argument shows, in fact, that if $f \in P^o$, then $f(e_n) = 0$ for all but a finite number of n).

<u>1.9.2</u>. Suppose that P is a cone with a base B. If $a_i, b_j \in B$ and

$$\sum_{i=1}^{m} \lambda_i a_i \leq \sum_{j=1}^{n} \mu_j b_j \text{ , then } \sum_{i=1}^{m} \lambda_i \leq \sum_{j=1}^{n} \mu_j .$$

Equality signs can replace the inequality signs here.

<u>Proof</u>. Apply the linear functional defined in 1.9.1.

<u>1.9.3. COROLLARY</u> Suppose that B is a base for P. If $b_1, b_2 \in B$ and $b_1 \leq b_2$, then $b_1 = b_2$.

<u>Proof</u> If f is the linear furctional defined in 1.9.1, then $f(b_2 - b_1) = 0$.

<u>1.9.4. COROLLARY</u>. If B is a base for the cone P, and K a convex set of positive real numbers, then $\bigcup\{\lambda B : \lambda \in K\}$ is convex.

<u>Proof</u>. Suppose that $b_i \in B$ and $\lambda_i \in K$ (i = 1,2). Take μ in (0.1), and let $x = \mu\lambda_1 b_1 + (1 - \mu)\lambda_2 b_2$. Then $x \in P$, so $x = \nu b$ for some $\nu > 0$ and b in B. By 1.9.2, $\nu = \mu\lambda_1 + (1 - \mu)\lambda_2 \in K$.

In particular, the set $C = \bigcup\{\lambda B : 0 \leq \lambda \leq 1\}$ has the property that both C and $P \sim C$ are convex. This has led to a generalisation of the idea of a base, called a <u>cap</u> (see Phelps (1) and section 3.12).

Clearly, if E is the kernel of a strictly monotonic linear functional, then $E \cap P = \{0\}$. The converse also holds:

<u>1.9.5</u>. Let P be a cone in a linear space X, and let E be the kernel of an element f of X'. If $E \cap P = \{0\}$, then either f or -f is strictly monotonic.

<u>Proof</u>. If neither f nor -f is strictly monotonic, then neither is monotonic, and there exist x,y in P such that f(x) = 1, f(y) = -1.

Then $x + y \in E \cap P$, so $x + y = 0$. But this contradicts the assumption that P is a cone.

Hence P has a base iff there is a maximal subspace E such that $E \cap P = \{0\}$.

The next result can easily be made more general, but we will only require it in the form given here.

<u>1.9.6</u>. Suppose that B is a base for the cone P. Then an element of B is an extreme point of B iff it is an extremal point of P.

<u>Proof</u>. Suppose that x is an extreme point of B. If $x = y + z$, where $y,z \in P \sim \{0\}$, then there exist a,b in B and $\lambda,\mu > 0$ such that $y = \lambda a$, $z = \mu b$. By 1.9.2, $\lambda + \mu = 1$. Hence $a = b = x$, so $y = \lambda x$.

Now suppose that x is an extremal point of P, and that $x = \lambda a + (1 - \lambda)b$, where $a,b \in B$ and $0 < \lambda < 1$. Then $0 \leqslant \lambda a \leqslant x$, so $\lambda a = \mu x$ for some $\mu > 0$. Since a and x are in B, $\mu = \lambda$ and $a = x$.

<u>1.9.7</u>. Suppose that P is an Archimedean cone containing at least two linearly independent elements, and that B is a base for P. If $0 \leqslant x \leqslant B$, then $x = 0$.

<u>Proof</u>. There exist b_0 in B and $\lambda \geqslant 0$ such that $x = \lambda b_0$. Let $\lambda_0 = \sup \{\lambda \geqslant 0 : \lambda b_0 \leqslant B\}$. Since P is Archimedean, $\lambda_0 b_0 \leqslant B$. If $\lambda_0 \geqslant 1$, this implies that $B = \{b_0\}$, by 1.9.3. Hence $\lambda_0 < 1$. Take b in B. Then $b > \lambda_0 b_0$, so there exist $b' \in B$ and $\mu > 0$ such that $b = \lambda_0 b_0 + \mu b'$.

By 1.9.2, $\mu = 1 - \lambda_0$. But $b' \geqslant \lambda_0 b_0$, so $b \geqslant \lambda_0(2 - \lambda_0)b_0$. This is true for all b in B, so $\lambda_0 = 0$.

This result is not true if P is not Archimedean. For example, let P be the cone $\{0\} \cup \{(\xi,\eta) : \xi > 0\}$ in R^2. A base for P is $B = \{(\xi,\eta) : \xi = 1\}$, and $(\xi,0) \leqslant B$ for $0 < \xi < 1$.

1.10. Regular and maximal order-convex subspaces

The guiding principle of this section is the analogy between order-convex subspaces and ideals of a ring. The results are not needed for later chapters, but are of some interest in themselves.

1.10.1. Let X be a linear space with an antisymmetric ordering. If X has no proper order-convex subspaces, then the dimension of X is 0 or 1.

Proof. We assume that $X \neq \{0\}$. If there is an element x not in $P \cup (-P)$, then lin x is order-convex, so is the whole space. Suppose now that $P \cup (-P) = X$. Take $e > 0$. Then $[\text{lin } e] = X$, so $e \in P_0$. Hence X is one-dimensional, by 1.5.11.

1.10.2. Let E be a maximal member of the family of proper order-convex subspaces of X. Then E is a maximal proper subspace, and there is a monotonic linear functional with kernel E.

Proof. The usual reasoning shows that X/E has no proper order-convex subspaces. Since the ordering in X/E is antisymmetric (1.4.4), E is a maximal proper subspace, by 1.10.1. The rest follows by 1.5.5.

By a "maximal subspace" we shall mean a maximal proper subspace. We say that a proper order-convex subspace is _regular_ if there is an order-unit in X/E. By 1.4.4, this is equivalent to the existence of an element a in $X \sim E$ such that $[E + \text{lin } a] = X$. (This definition and the theory below are due to Bonsall (1)). An order-convex maximal subspace is clearly regular.

1.10.3. Every regular order-convex subspace is contained in an order-convex maximal subspace.

Proof. Suppose that $[\, E + \text{lin } a \,] = X$, and let F be the family of all proper order-convex subspaces containing E. Then no member of F contains a, so the union of a nest in F is itself F. Zorn's lemma and 1.10.2 now give the result.

If there are order-units in X, then every proper order-convex subspace is regular. Furthermore, we have:

<u>1.10.4</u>. Suppose that there are order-units in X, and that E is a subspace containing no order-unit. Then E is contained in an order-convex maximal subspace.

<u>Proof</u>. If e is an order-unit and $y \geq e$, then y is an order-unit. Hence [E] contains no order-unit, so is a regular order-convex subspace.

Putting $E = \{0\}$, we see that this gives an alternative proof of 1.6.3 (and hence of the separation and extension theorems).

<u>1.10.5</u>. If P is a proper wedge such that $P \cup (-P) = X$, and P is not everywhere non-Archimedean, then X has order-convex maximal subspaces.

<u>Proof</u>. Define a subspace K of X by: $x \in K$ iff there exists a in P such that $-a \leq nx \leq a$ for all n in ω. Then K is order-convex, and $K \subset X$, since X is not everywhere non-Archimedean. Let r be the projection on X onto X/K. It is clear that $r(P) \cup r(-P) = X/K$. We show that r(P) is almost Archimedean, from which it follows that X/K is one-dimensional, by 1.5.11. Suppose that $a \notin K$ and $-r(a) \leq nr(x) \leq r(a)$ for all n in ω. If $n_0 x \geq a$ for some n_0 in ω, then $n_0 r(x) = r(a)$, so $2r(a) = 2n_0 r(x) \leq r(a)$, which is a contradiction. Similarly, we cannot have $n_0 x \leq -a$ for any n_0 in ω, so in fact $-a \leq nx \leq a$ for all n in ω, and $x \in K$, i.e. $r(x) = 0$, as required.

<u>1.10.6</u>. If E is an order-convex subspace of an ordered linear space X, then one of the following is true:

(1) E is contained in an order-convex maximal subspace;

(2) there is a subspace K containing E such that X/K is totally ordered and everywhere non-Archimedean.

Proof. If $P \subseteq E$, then any subspace containing E is order-convex, so (1) occurs.

Suppose that there is an element u in $P \sim E$. By Zorn's lemma, there is a maximal member K of the family of order-convex subspaces containing E but not u. If $x \notin K$, then $u \in [K + \text{lin } x]$, i.e. there exist k_1 in K and λ_1 in R such that

$$k_1 + \lambda_1 x \leq u \leq k_2 + \lambda_2 x.$$

If $\lambda_1 = \lambda_2$, then $u - \lambda_1 x \in K$, since K is order convex, and $\lambda_1 \neq 0$, since $u \notin K$. If $\lambda_1 \neq \lambda_2$, then $(\lambda_2 - \lambda_1)x \geq k_1 - k_2$. In either case, $r(x) \in r(P) \cup r(-P)$. Hence X/K is totally ordered. If X/K is not everywhere non-Archimedean, then it has an order-convex maximal subspace, by 1.10.5, from which it follows that E is contained in an order-convex maximal subspace of X.

CHAPTER 2

LINEAR LATTICES AND RIESZ SPACES

We recall the hierarchy of interpolation properties described in Section 0.1. By a linear lattice we mean an ordered linear space in which the ordering is a lattice ordering, and by a Riesz space we mean a linear space with a directed, antisymmetric ordering satisfying the Riesz interpolation property. Commutative lattice groups and Riesz groups are defined similarly. (In the Russian literature, a linear lattice is called a K-lineal. Some writers use the term "Riesz space" to mean "linear lattice", but our usage is justified by the fact that what we are calling a Riesz space was defined and studied by Riesz (1).)

The first five sections of this chapter are concerned with Riesz spaces, linear lattices and order-complete spaces, in that order. Almost everything in these sections applies to commutative groups. The basic theory of linear lattices (or commutative lattice groups) consists of a large collection of elementary results (2.2, 2.3). In 2.4, we introduce bands and annihilators; $A^{\perp}$ is always a band, and in an Archimedean lattice, every band is of this form. If A is a band in an order-complete space X, then X is the ordered direct sum of A and $A^{\perp}$ (2.5.4). Linear mappings with values in order-complete spaces share many of the properties of linear functionals; in particular, the Hahn-Banach and monotone extension theorems apply.

In sections 2.6 and 2.7, we consider linear mappings defined on Riesz spaces and linear lattices. The most important result is the Riesz decomposition theorem (2.6.1): if X is a Riesz space, then X' is an order-complete linear lattice, and it is easy to give explicit formulae describing the lattice operations in X'. Like other results, this can be extended (with no change of proof) to linear mappings with values in an order-complete space. An extension theorem of Hahn-Banach type applies to monotonic linear mappings dominated by a sub-linear mapping p such that $|x_1| \leq |x_2|$ implies $p(x_1) \leq p(x_2)$ (2.6.3). The extremal monotonic linear functionals on a linear lattice are precisely the linear lattice homomorphisms into R: to obtain corresponding results for linear mappings, we introduce the class of discrete linear mappings (2.7).

For more comprehensive studies of linear lattices (especially order-complete ones), we refer to the books of Luxemburg-Zaanen (1), Vulih (1) and Nakano (1).

2.1. Riesz spaces

Our first result gives two possible alternative definitions of a Riesz group:

2.1.1. THEOREM. Let X be a commutative group with a directed, anti-symmetric ordering. Then the following statements are equivalent:

(i) X is a Riesz group;

(ii) if x_1, x_2, y are positive elements such that $0 \leq y \leq x_1 + x_2$, then there exist y_i such that $0 \leq y_i \leq x_i$ $(i = 1,2)$ and $y = y_1 + y_2$.

(iii) if $a,b,c,d \geq 0$ and $a + b = c + d$, then there exist $p,q,r,s \geq 0$ such that $a = p + q$, $b = r + s$, $c = p + r$, $d = q + s$.

Proof. (i) => (ii). If x_1,x_2,y are as given, then each of x_1,y is greater than each of 0, $y - x_2$, so there is an intermediate element y_1. Then $0 \leq y_1 \leq x_1$, and $y - x_2 \leq y_1 \leq y$, so $0 \leq y - y_1 \leq x_2$.

(ii) => (iii). With a,b,c,d as given, we have $0 \leq a \leq c + d$. Therefore (ii) says that there exist p,q such that $0 \leq p \leq c$, $0 \leq q \leq d$ and $a = p + q$. Let $r = c - p$, $s = d - q$. Then $r,s \geq 0$, and $b = c + d - a = r + s$.

(iii) => (i). Suppose that $a_i \leq b_j$ for $i,j = 1,2$.

Then we have

$$(b_1 - a_1) + (b_2 - a_2) = (b_1 - a_2) + (b_2 - a_1),$$

and each of these four elements is positive. By (iii), there exist $p,q,r,s \geq 0$ such that $b_1 - a_1 = p + q$, $b_1 - a_2 = p + r$, $b_2 - a_1 = q + s$. Then $a_1 + q$ is the required intermediate element, since $a_1 + q = b_2 - s$, and also $a_1 + q = b_1 - p = a_2 + r$.

<u>2.1.2. COROLLARY</u>. If X is a Riesz group, and $x_1, \ldots, x_n$ y are positive elements such that $0 \leq y \leq x_1 + \ldots + x_n$, then there exist y_i such that $0 \leq y_i \leq x_i$ $(i = 1, \ldots, n)$ and $y = y_1 + \ldots + y_n$.

<u>Proof</u>. Induction.

Statement (iii) of 2.1.1 can also be extended to finite collections of elements.

<u>2.1.3</u>. If a_i, b_i are elements of a Riesz group such that $a_i \leq b_i$ $(i = 1,2)$, then

$$[a_1, b_1] + [a_2, b_2] = [a_1 + a_2, b_1 + b_2].$$

<u>Proof</u>. We saw in 1.1.5 that the right-hand side contains the left hand side. Suppose that $a_1 + a_2 \leq x \leq b_1 + b_2$. Let $y = x - (a_1 + a_2)$. Then $0 \leq y \leq (b_1 - a_1) + (b_2 - a_2)$, so there exist y_1, y_2 such that $0 \leq y_i \leq b_i - a_i$ $(i = 1,2)$ and $y = y_1 + y_2$. Then $y_i + a_i \in [a_i, b_i]$ $(i = 1,2)$, and $x = (y_1 + a_1) + (y_2 + a_2)$.

It is clear that any linear lattice is a Riesz space. A simple example of a Riesz space that is not a linear lattice is R^2, ordered by the cone $\{0\} \cup \{(\xi,\eta) : \xi,\eta > 0\}$. An example of an Archimedean Riesz space that is not a lattice is given in Namioka (1), p.45.

In Riesz groups, the converse of 1.4.2 applies:

<u>2.1.4</u>. Let X be a Riesz group, and let E_1, E_2 be order-convex,

directed subgroups whose direct sum is X. Then X is the ordered direct sum of E_1 and E_2.

<u>Proof</u>. Suppose that $x_1 + x_2 \geq 0$, where $x_i \in E_i$. There exist u_i, v_i in $E_i \cap P$ such that $x_i = u_i - v_i$. Then $u_1 + u_2 \geq v_1 + v_2 \geq 0$, so there exist w_1, w_2 such that $0 \leq w_i \leq u_i$ and $v_1 = w_1 + w_2$. But then $w_2 \in E_1 \cap E_2$, since E_1 and E_2 are order-convex, so $w_2 = 0$. Hence $x_1 = u_1 - w_1 \geq 0$.

<u>2.2. Linear lattices: basic theory</u>

We start by showing that rather weaker assumptions are sufficient to ensure that an ordered commutative group is a lattice:

<u>2.2.1</u>. If X is a commutative group with an antisymmetric ordering, and $\sup\{x,0\}$ exists for each x in X, then X is a lattice.

<u>Proof</u>. For any x,y, we know that $\sup\{x - y, 0\}$ exists. By 1.1.7, it follows that $\sup\{x,y\}$ exists. By 1.1.8, $\inf\{x,y\}$ also exists for all x,y.

We use the usual notation $x \vee y$, $x \wedge y$ for $\sup\{x,y\}$ and $\inf\{x,y\}$. From 1.1.7 and 1.1.8, we have two simple equalities which will be used repeatedly:

<u>2.2.2</u>. If x,y,z are elements of a commutative lattice group, then

(i) $x \vee y + z = (x + z) \vee (y + z)$,

(ii) $(-x) \vee (-y) = -(x \wedge y)$.

In a linear lattice, it is clear that $\lambda(x \vee y) = (\lambda x) \vee (\lambda y)$ and $\lambda(x \wedge y) = (\lambda x) \wedge (\lambda y)$ for all $\lambda > 0$. However, in a commutative lattice group, these relations do not necessarily hold for positive integers λ.

<u>2.2.3. THEOREM</u>. If x,y are elements of a commutative lattice group, then

$$x \vee y + x \wedge y = x + y.$$

<u>Proof</u>. We have

$$\begin{aligned} x \vee y &= (-y + (x + y)) \vee (-x + (x + y)) \\ &= (-y) \vee (-x) + (x + y), \quad \text{by 2.2.2(i)}, \\ &= -(x \wedge y) + (x + y), \quad \text{by 2.2.2(ii)}. \end{aligned}$$

An obvious result which is useful in providing examples is the following:

<u>2.2.4</u> Suppose that X,Y are commutative groups, and that Q is the positive set for a lattice ordering in Y. If f is a group isomorphism on X onto Y, then $f^{-1}(Q)$ gives a lattice ordering in X. For x_1, x_2 in X,

$$x_1 \vee x_2 = f^{-1}(f(x_1) \vee f(x_2)).$$

The next result is a useful application of 2.1.1.

<u>2.2.5</u>. If x,y,z are positive elements of a commutative lattice group, then

$$x \wedge (y + z) \leq (x \wedge y) + (x \wedge z).$$

<u>Proof</u>. Let $x \wedge (y + z) = u$. By 2.1.1, there exist y_1, z_1 such that $0 \leq y_1 \leq y$, $0 \leq z_1 \leq z$ and $y_1 + z_1 = u$. Now $y_1 \leq u \leq x$, so $y_1 \leq x \wedge y$. Similarly, $z_1 \leq x \wedge z$.

<u>2.2.6</u>. The following condition is sufficient for a linear lattice to be Archimedean: if $x,y \in P$ and $nx \leq y$ $(n \in \omega)$, then $x = 0$. An almost Archimedean linear lattice is Archimedean.

<u>Proof</u>. If X is not Archimedean, then there exist x,y such that $x \notin -P$, $y \in P$ and $nx \leq y$ $(n \in \omega)$. Let $x' = x \vee 0$. Then $x' > 0$, and $nx' = (nx) \vee 0 \leq y$ $(n \in \omega)$.

Distributivity

If A,B are subsets of a lattice, we shall write $A \vee B$ to denote the set $\{a \vee b : a \in A, b \in B\}$, and $\vee A$ to denote the set of suprema of finite subsets of A (with similar conventions for $\wedge$).

2.2.7. THEOREM. Let A,B be subsets of a commutative lattice group.

(i) If A,B have suprema a_0, b_0, then $a_0 \wedge b_0 = \sup (A \wedge B)$.

(ii) If A,B have infima a_1, b_1, then $a_1 \vee b_1 = \inf (A \vee B)$.

In particular, a commutative lattice group is distributive.

Proof. (i) We show first that, for any b in X,

$$a_0 \wedge b = \sup (A \wedge b) \qquad (1).$$

Clearly, $a_0 \wedge b \geq A \wedge b$. If $x \geq A \wedge b$, then, for any a in A,

$$\begin{aligned} x + a_0 \vee b &\geq x + a \vee b \\ &\geq a \wedge b + a \vee b \\ &= a + b, \text{ by 2.2.3.} \end{aligned}$$

Hence $x + a_0 \vee b \geq a_0 + b$, so, using 2.2.3 again, $x \geq a_0 \wedge b$, as required.

Now suppose that $y \geq A \wedge B$. For each b in B, $y \geq A \wedge b$, so, by (1), $y \geq a_0 \wedge b$. Hence $y \geq a_0 \wedge B$, so, by (1) again, $y \geq a_0 \wedge b_0$. This completes the proof.

(ii) follows, by considering -A, -B.

One consequence of distributivity is the following finite extension of 2.2.3:

2.2.8. If $x_1, \ldots, x_n$ are elements of a commutative lattice group, then

$$x_1 \vee \ldots \vee x_n = \sum_{k=1}^{n} (-1)^{k-1} \sum_{\sigma \in R_{k,n}} x_{\sigma(1)} \wedge \ldots \wedge x_{\sigma(k)},$$

where $R_{k,n}$ is the set of mappings σ on $\{1, \dots, k\}$ into $\{1,\dots,n\}$ such that $\sigma(i) < \sigma(j)$ whenever $i < j$.

Proof. The result is true for $n = 2$, by 2.2.3. Suppose that it is true for some positive integer n, and take $x_1,\dots,x_{n+1}$ in X. Let $y = x_1 \vee \dots \vee x_n$. Then, by 2.2.3,

$$x_1 \vee \dots \vee x_{n+1} = y + x_{n+1} - y \wedge x_{n+1}.$$

By 2.2.7,

$$y \wedge x_{n+1} = (x_1 \wedge x_{n+1}) \vee \dots \vee (x_n \wedge x_{n+1}).$$

Applying the induction hypothesis to this expression and to y, we obtain the desired expression for $x_1 \vee \dots \vee x_{n+1}$.

Sublattices

A sublattice of a lattice is a subset admitting $\vee$ and $\wedge$. An intersection of sublattices is a sublattice, so there is a smallest sublattice containing a given set. In a distributive lattice, the smallest sublattice containing a subset A is $\vee(\wedge A) = \wedge(\vee A)$.

The argument used in 2.2.1 shows that a subgroup E of a commutative la tice group is a sublattice iff $x \in E$ implies $x \vee 0 \in E$. Clearly, such a subgroup is directed. Conversely, it is obvious that an order-convex, directed subgroup is a sublattice.

The usual ordering of R^3 is a lattice ordering. The set of elements of the form $(\xi, \eta, \xi+\eta)$ is a directed linear subspace, but not a sublattice. We notice that it is the sum of two linear sublattices, namely the one-dimensional subspaces generated by $(1,0,1)$ and $(0,1,1)$. (It is also a linear lattice under its own ordering, but the lattice operations do not agree with those on R^3.)

We now give a few permanence properties that will be required:

2.2.9. If A is a sublattice of a commutative lattice group, then so is $[A]$.

Proof. Take x_1, x_2 in $[A]$. There exist a_i, b_i in A such that

$a_i \leq x_i \leq b_i$ $(i = 1,2)$. Then $x_1 \vee x_2$ and $x_1 \wedge x_2$ are both in $[a_1 \wedge a_2, b_1 \vee b_2]$, which is contained in A.

2.2.10. If A is a convex subset of a linear lattice, then $\vee A$, $\wedge A$ and $\vee(\wedge A)$ are convex.

Proof. Take x_1, x_2 in $\vee A$. Then $x_i = \sup A_i$, where A_i is a finite subset of A $(i = 1,2)$. If $0 < \lambda < 1$, then

$$\begin{aligned}\lambda x_1 + (1-\lambda)x_2 &= \sup(\lambda A_1) + \sup(1-\lambda)A_2 \\ &= \sup(\lambda A_1 + (1-\lambda)A_2), \text{ by 1.1.7.}\end{aligned}$$

Hence $\lambda x_1 + (1-\lambda)x_2 \in \vee A$.

2.2.11. If A is a linear subspace of a linear lattice, then the sublattice generated by A is a linear subspace.

Proof. By an argument similar to that used in 2.2.10, it is easily seen that $\vee A$ and $\wedge(\vee A)$ admit addition. It is also clear that they admit multiplication by positive scalars. If $x \in \wedge(\vee A)$, then there exist y_i in $\vee A$ such that $x = y_1 \wedge \dots \wedge y_n$. Then $-y_i \in \wedge A$, so $-x = (-y_1) \vee \dots \vee (-y_n) \in \vee(\wedge A)$.

The following slightly clumsy lemma on upper bounds will be useful later. By a "lattice subgroup" we mean a subgroup that is also a sublattice.

2.2.12. Let E be a lattice subgroup of a commutative lattice group X, and let A be a subset of E that has upper bounds in X. Then there exist a_0 in A and a subset A' of $E \cap P$ such that A' admits $\vee$ and that x is an upper bound of A' iff $x + a_0$ is an upper bound of A.

Proof. Take a_0 in A, and let $A_1 = \vee(A - a_0)$. Then A_1 contains 0 and admits $\vee$, and $A_1 \subseteq E$, since E is a sublattice. Also, $x \geq A_1$ iff $x + a_0 \geq A$. Let $A' = A_1 \cap P$. Then A' admits $\vee$, and if $b \in A_1$, then $b \vee 0 \in A'$, so A' has the same upper bounds as A_1.

Examples

(i) The space of all real-valued functions on a set S is a lattice under its usual ordering. The lattice operations are defined pointwise, viz:

$$(x \vee y)(s) = x(s) \vee y(s) \quad (s \in S),$$

where the second occurrence of $\vee$ refers to the usual ordering of R. Taking S finite or countable, we obtain R^n or the space of all real sequences. In the latter case, the sequence spaces m, c_o, l_1, F are sublattices.

(ii) Any totally ordered linear space is a linear lattice.

(iii) Let the space of all real polynomials be ordered as functions on $[0,1]$ (cf. section 1.1, ex. (v)). We show that this is not a lattice. Let $f(t) = t$, $g(t) = 1-t$. If h is a polynomial and $h \geq f$, $h \geq g$, then $h(\frac{1}{2}) > \frac{1}{2}$. By Weierstrass's theorem, there is a polynomial h' such that $h' \geq f$, $h' \geq g$ and $h'(\frac{1}{2}) < h(\frac{1}{2})$.

(iv) As in section 1.1, let P_s denote the set of sequences (within the space considered) having all partial sums non-negative. We use 2.2.4 to show that P_s gives a lattice ordering of the space s of all real sequences. An isomorphism f of s onto itself is defined by $f(x) = \{X_n\}$, where $x = \{\xi_n\}$, $X_n = \xi_1 + \dots + \xi_n$. If P denotes the set of all non-negative sequences (i.e. the "usual" positive cone), then $f^{-1}(P) = P_s$. The supremum (with respect to the P_s-ordering) of two elements $x = \{\xi_n\}$, $y = \{\eta_n\}$ is $\{\zeta_n\}$, where $Z_n = X_n \vee Y_n$ $(n \geq 1)$, $\zeta_n = Z_n - Z_{n-1}$ $(n \geq 2)$ and $\zeta_1 = Z_1$.

If λ, μ are real numbers, it is elementary that

$$|\lambda \vee \mu - \lambda' \vee \mu'| \leq |\lambda - \lambda'| \vee |\mu - \mu'|.$$

It follows that (with the above notation), $|\zeta_n| \leq |\xi_n| \vee |\eta_n|$ for all n, so that the spaces m, c_o, l_1, F are sublattices of s with the P_s- ordering. We notice, however, that m is not order-convex in s with respect to this ordering, since

$$(1,-1,2,-2,3,-3, \ldots) + (0,1,-1,2,-2,3,\ldots) = (1,0,1,0,1,0,\ldots) \in m.$$

(v) Let P_d denote the set of decreasing, non-negative sequences within the space considered. It is elementary that P_d generates F, but not c_o or l_1. Now a one-to-one linear mapping f on l_1 into c_o is defined by $f\{\xi_n\} = \{\eta_n\}$, where $\eta_n = \sum_{r=n}^{\infty} \xi_r$. The inverse mapping is obtained from the relation $\xi_n = \eta_n - \eta_{n+1}$ $(n \geq 1)$. If P denotes the set of non-negative sequences in l_1, then $f(P) = P_d$. Hence the range of f is $P_d - P_d$ (= Y, say), and (applying 2.2.4 to f^{-1}) P_d gives Y a lattice ordering. If $y = \{\eta_n\} \in Y$, then the supremum of y and 0 with respect to the P_d- ordering is $\{\zeta_n\}$, where

$$\zeta_n = \sum_{r=n}^{\infty} (\eta_r - \eta_{r+1}) \vee 0.$$

The space F is clearly a sublattice (isomorphic to F with its usual ordering).

<u>Quotient spaces</u>

<u>2.2.13</u>. Let X be a commutative lattice group, and let E be a subgroup of X that is an order-convex sublattice. Then X/E is a lattice, and for x,y in X, we have $r(x \vee y) = r(x) \vee r(y)$, where r is the projection of X onto X/E.

<u>Proof</u>. The quotient ordering is antisymmetric, by 1.4.4(iv). Let $z = x \vee y$. Then $r(z) \geq r(x)$ and $r(z) \geq r(y)$. Suppose that $r(u) \geq r(x)$ and $r(u) \geq r(y)$. Then there exist e,f in E such that $u + e \geq x$ and $u + f \geq y$. Then $u + e \vee f \geq z$, so $r(u) \geq r(z)$. Hence $r(z) = \sup \{r(x), r(y)\}$.

<u>Note</u> It is possible for X/E to be a lattice when E is not a sublattice of X. For instance, if X is R^2 with the usual ordering, and E is the set of elements of the form $(\xi, -\xi)$, then X/E is isomorphic to R with the usual ordering. However, it is clear that if X/E is a lattice and r is a lattice homomorphism, then E must be a sublattice.

Linear lattices with bases for the positive cone

<u>2.2.14</u>. If the positive cone in a linear lattice X has a base B, then the following statements hold:

(i) The ordering is Archimedean.

(ii) If the dimension of X is at least two, then $\inf B = 0$.

(iii) If a,b are distinct extreme points of B, then $a \wedge b = 0$.

<u>Proof</u>. (i) Suppose that $x,y \geq 0$ and $nx \leq y$ ($n \in \omega$). There exist $\lambda,\mu \geq 0$ and a,b in B such that $x = \lambda a$, $y = \mu b$. By 1.9.2, $n\lambda \leq \mu$ ($n \in \omega$). Hence $\lambda = 0$, so $x = 0$, and the ordering is Archimedean, by 2.2.6.

(ii) Suppose that $x \leq B$. Then $0 \leq x \vee 0 \leq B$. Since P generates X, B contains at least two linearly independent elements. Since P is also Archimedean, 1.9.6 applies to show that $x \vee 0 = 0$, or $x \leq 0$.

(iii) By 1.9.6, a and b are extremal points of P, so there exist $\lambda,\mu \geq 0$ such that $a \wedge b = \lambda a = \mu b$. If $\lambda,\mu > 0$, this implies that $a = b$.

A characterisation of convex sets that are bases for cones giving a lattice ordering is given by Kendall (1).

A note on commutative lattice groups

It follows from the definition of an ordered linear space that if $nx \geq 0$ for some positive integer n, then $x \geq 0$. However, this need not be true in an ordered commutative group: for instance, let the integers be ordered by the semigroup consisting of all the non-negative integers except 1. It is of interest that this property does hold in lattice-ordered groups. To prove this, we need the following result:

<u>2.2.15</u>. If X is a commutative lattice group, $x \in X$ and n is a positive integer, then

$$n(x \wedge 0) = (nx) \wedge (n-1)x \wedge \ldots \wedge x \wedge 0.$$

<u>Proof</u>. The statement is true for $n = 1$. Suppose that it is true for $n = r$. Write

$$y = r(x \wedge 0)$$
$$= (rx) \wedge (r-1)x \wedge \ldots \wedge x \wedge 0.$$

Then

$$(r+1)(x \wedge 0) = x \wedge 0 + y$$
$$= (x + y) \wedge y, \text{ by } 2.2.2(i).$$

Now

$$x + y = (r + 1)x \wedge (rx) \wedge \ldots \wedge 2x \wedge x,$$

so

$$(r + 1)(x \wedge 0) = (r + 1)x \wedge (rx) \wedge \ldots \wedge x \wedge 0,$$

as required.

<u>2.2.16. COROLLARY</u>. If X is a commutative lattice group, $x \in X$ and $nx \geqslant 0$ for some positive integer n, then $x \geqslant 0$.

<u>Proof</u>. We have $(nx) \wedge 0 = 0$, so, by 2.2.15,

$$n(x \wedge 0) = (n - 1)x \wedge \ldots \wedge x \wedge 0$$
$$= (n - 1)(x \wedge 0).$$

Hence $x \wedge 0 = 0$, so $x \geqslant 0$.

<u>2.3. Positive and negative parts and moduli</u>

In a commutative lattice group, we define:

$x^+ = x \vee 0$ (the <u>positive part</u> of x)
$x^- = (-x) \vee 0$ (the <u>negative part</u> of x)

$$|x| = x \vee (-x) \qquad \text{(the \underline{modulus} of } x).$$

Each of these is positive (in the case of $|x|$, this follows from 2.2.16), and $-x^- = x \wedge 0$, so by 2.2.3 we have $x = x^+ - x^-$. Furthermore, this is a minimal decomposition of x in the sense that if $x = u - v$, where $u, v \geq 0$, then $u \geq x^+$ and $v \geq x^-$. We notice that $|x| \leq a$ iff $\pm x \leq a$, i.e. $-a \leq x \leq a$, and that this implies that $x^+ \leq a$ and $x^- \leq a$.

In the space of all real-valued functions on a set S, the operations $x^+, |x|$ are pointwise extensions of the corresponding operations on R: $x^+(s) = (x(s))^+$ and $|x|(s) = |x(s)|$ $(s \in S)$.

We now give a number of elementary relations satisfied by these quantities. Unless otherwise stated, the elements concerned are assumed to be elements of a commutative lattice group.

<u>2.3.1</u>. (i) $x^+ \wedge x^- = 0$.

(ii) $|x| = x^+ \vee x^- = x^+ + x^-$.

(iii) If $x = u - v$ and $u \wedge v = 0$, then $u = x^+$, $v = x^-$.

<u>Proof</u>. (i) Since $x = x^+ - x^-$, we have

$$(x^+ \wedge x^-) - x^- = x \wedge 0 \quad = -x^-.$$

(ii) We have $x^+ \vee x^- = x \vee (-x) \vee 0$. This is equal to $x \vee (-x)$, since $x \vee (-x) \geq 0$. By (i) and 2.2.3, $x^+ + x^- = x^+ \vee x^-$.

(iii) Under these hypotheses, $u \geq 0$ and $u \geq x$, so $u \geq x^+$. If $y \geq 0$ and $y \geq x$, then $u - y \leq u$ and $u - y \leq v$, so $u - y \leq 0$. Hence $u = x^+$.

From (ii), we see that $2x^+ = |x| + x$, $2x^- = |x| - x$. It follows that, in a linear lattice, a linear subspace admitting moduli is a sublattice. This is not true for subgroups, as is shown by the example $\{(m,n) : m + n \text{ even}\}$ in R^2.

<u>2.3.2</u>. (i) $(x + y)^+ \leq x^+ + y^+$; $(x + y)^- \leq x^- - y^-$.

(ii) $|x + y| \leq |x| + |y|$.

(iii) In a linear lattice, $|\lambda x| = |\lambda| . |x|$ $(\lambda \in R)$.

<u>Proof</u>. Immediate.

For differences of two elements, we have a long string of useful equalities and inequalities.

<u>2.3.3</u>. (i) $(x - y)^+ = x \vee y - y = x - x \wedge y$.

(ii) $(x - y)^- = x \vee y - x = y - x \wedge y$.

(iii) $|x - y| = x \vee y - x \wedge y$.

<u>Proof</u>. By 2.2.2(i), $(x - y)^+ = x \vee y - y$. By 2.2.3, this is equal to $x - x \wedge y$. This proves (i). (ii) can be proved similarly, and (iii) follows by addition.

Addition of 2.3.3(iii) and 2.2.3 gives:

<u>2.3.4 COROLLARY</u>.

(i) $2(x \vee y) = (x + y) + |x - y|$.

(ii) $2(x \wedge y) = (x + y) - |x - y|$.

<u>2.3.5</u>. (i) $x^+ - y^+ \leq (x - y)^+$; $x^- - y^- \leq (x - y)^-$.

(ii) $|x^+ - y^+| \leq |x - y|$; $|x^- - y^-| \leq |x - y|$.

(iii) $||x| - |y|| \leq |x - y|$.

<u>Proof</u>. (i) By 2.3.2(i), $(x - y)^+ + y^+ \geq x^+$.

(ii) By (i), $x^+ - y^+ \leq |x - y|$. Hence also $y^+ - x^+ \leq |y - x| = |x - y|$.

(iii) By 2.3.2(ii), $|x| - |y| \leq |x - y|$. Hence also $|y| - |x| \leq |y - x| = |x - y|$.

Some of the above inequalities are sharpened by the following equalities:

<u>2.3.6</u>. (i) $|x + y| \vee |x - y| = |x| + |y|$.

(ii) $|x + y| \wedge |x - y| = ||x| - |y||$.

<u>Proof</u>. (i) By definition,

$$|x + y| \vee |x - y| = (x + y) \vee (-x-y) \vee (x - y) \vee (-x+y).$$

By 2.2.2(i),

$$(x + y) \vee (-x+y) = |x| + y,$$

$$(-x-y) \vee (x - y) = |x| - y,$$

so (again by 2.2.2(i)),

$$|x + y| \vee |x - y| = |x| + |y|.$$

(ii) By distributivity,

$$(x + y) \wedge |x - y| = ((x + y) \wedge (x - y)) \vee ((x+y) \wedge (y-x))$$
$$= (x - |y|) \vee (y - |x|).$$

Applying this to $-x,-y$, we have also:

$$(-x-y) \wedge |x - y| = (-x - |y|) \vee (-y - |x|).$$

Therefore, by 2.2.2(i),

$$|x + y| \wedge |x - y| = (|x| - |y|) \vee (|y| - |x|).$$

<u>2.3.7. COROLLARY</u>.

(i) $|u + v| + |u - v| = 2(|u| \vee |v|)$.

(ii) $||u + v| - |u - v|| = 2(|u| \wedge |v|)$.

(iii) $|u| \wedge |v| = 0 \iff |u + v| = |u - v|$.

<u>Proof</u>. Let $x = u + v$, $y = u - v$.

<u>2.3.8</u>. $|x \vee y| \leq |x| \vee |y|$; $|x \wedge y| \leq |x| \vee |y|$.

<u>Proof</u>. We have $x \vee y \leq |x| \vee |y|$, and $-(x \wedge y) \leq -x \leq |x| \vee |y|$. This proves the first inequality, and the second one follows.

The next result is a useful variant of 2.1.1.

<u>2.3.9</u>. If $|y| \leq x_1 + x_2$, where $x_1, x_2 \geq 0$, then there exist y_i such that $|y_i| \leq x_i$ $(i = 1,2)$ and $y = y_1 + y_2$.

Proof. By 2.1.1, there exist u_i, v_i such that $0 \leq u_i \leq x_i$, $0 \leq v_i \leq x_i$ $(i = 1,2)$ and $y^+ = u_1 + u_2$, $y^- = v_1 + v_2$. Let $y_i = u_i - v_i$. Then $\pm y_i \leq x_i$, so $|y_i| \leq x_i$, and $y = y_1 + y_2$.

Solid sets

A subset A of a commutative lattice group is said to be solid if $a \in A$ and $|x| \leq |a|$ implies that $x \in A$.

If A is solid, it is clearly symmetric, and if $x \in A$ then $|x|$, x^+, $x^- \in A$. In a linear lattice, a solid set is circled, and hence a solid subgroup is necessarily a linear subspace. The union and intersection of a family of solid sets are solid, and hence a given set A has a smallest solid set containing it and a largest solid subset. These are, respectively,

$$S(A) = \cup \{S(a) : a \in A\},$$

$$T(A) = \{x ; S(x) \subseteq A\},$$

where

$$S(x) = \{ y : |y| \leq |x| \} = [-|x|, |x|].$$

We notice that it follows from 2.3.9 that $S(x + y) \subseteq S(x) + S(y)$.

2.3.10. Let A be a subset of a commutative lattice group, and consider the statements:

(i) A is an order-convex sublattice;

(ii) A is order-convex and admits moduli;

(iii) A is solid.

If A is symmetric, then (i) => (ii) => (iii). If A is a subgroup, then the three statements are equivalent.

Proof. Suppose that A is symmetric. If (i) holds, and $a \in A$, then $|a| = a \vee (-a) \in A$. If (ii) holds, $a \in A$ and $|x| \leq |a|$, then $\pm|a| \in A$ and $-|a| \leq x \leq |a|$, so $x \in A$.

Now suppose that A is a solid subgroup. If $a \in A$, then $a^+ \in A$, since $|a^+| = a^+ \leq |a|$. Hence A is a sublattice. If $a \in A$ and $0 \leq x \leq a$, then $x \in A$, so A is order-convex, by 1.2.1.

It is easily seen that the unit disc in R^2 is solid, but neither order-convex nor a sublattice.

The following facts are simple consequences of 2.3.9:

<u>2.3.11</u>. (i) If A,B are solid subsets of a commutative lattice group, then A + B is solid.

(ii) If A is a solid subset of a linear lattice, then co(A) and lin A are solid.

The behaviour of the solid cover and interior is a little undisciplined. If A is convex or a linear subspace, then it follows from 2.3.9 that T(A) has the same property. However, convexity of A does not imply convexity of S(A): an example showing this is the line segment in R^2 joining (-2,0) and (1,3). We give one result which will be required in Chapter 4:

<u>2.3.12</u>. If A is a symmetric sublattice of a linear lattice, then $S(A) = [A]$, and this set is a convex sublattice.

<u>Proof</u>. If $a \in A$, then $|a| = a \vee (-a) \in A$. Thus if $x \in S(A)$, then there exists a in A such that $-a \leq x \leq a$, so that $x \in [A]$. Conversely, suppose that $x \in [A]$. Then there exist a,b in A such that $a \leq x \leq b$. Then $|x| \leq |a| \vee |b| \in A$, so $x \in S(A)$.

Take x,y in S(A). By the above, there exist a,b in A such that $|x| \leq a$, $|y| \leq b$ Let $c = a \vee b$. Then $c \in A$, and $|x \vee y| \leq |x| \vee |y| \leq c$, so $x \vee y \in S(A)$. Similarly, $x \wedge y \in S(A)$. Also, for $0 < \lambda < 1$,

$$\begin{aligned} |\lambda x + (1-\lambda)y| &\leq \lambda|x| + (1-\lambda)|y| \\ &\leq \lambda c + (1-\lambda)c \\ &= c, \end{aligned}$$

so $\lambda x + (1-\lambda)y \in S(A)$.

<u>2.3.13</u>. If A is a linear subspace of a linear lattice, then S(A) is a linear subspace.

<u>Proof</u>. By 2.2.9 and 2.3.10, S(A) is the order-convex cover of the sublattice generated by A. This is a linear subspace, by 1.2.3 and 2.2.11.

<u>2.4. Disjointness and bands</u>

Two elements x,y of a commutative lattice group X are said to be (lattice-) <u>disjoint</u> if $|x| \wedge |y| = 0$. We have seen (2.3.9) that this is equivalent to the statement $|x + y| = |x - y|$. We call a subset A of X <u>pairwise disjoint</u> if any two distinct members of A are disjoint.

The following statement is an immediate corollary of 2.2.5:

<u>2.4.1</u>. If $|y| \wedge |x_i| = 0$ $(i = 1, \ldots, n)$, then $|y| \wedge |x_1 + \ldots + x_n| = 0$.

The elementary properties of finite pairwise disjoint sets are summarised in the next result.

<u>2.4.2</u>. Suppose that $x_1,\ldots,x_n$ are pairwise disjoint elements of a commutative lattice group. Let $x = x_1 + \ldots + x_n$. Then:

(i) $x^+ = x_1^+ + \ldots + x_n^+$; $x^- = x_1^- + \ldots + x_n^-$.

(ii) $|x| = |x_1| + \ldots + |x_n| = |x_1| \vee \ldots \vee |x_n|$.

(iii) If $x \geq 0$, then $x_i \geq 0$ for each i. If $x = 0$, then $x_i = 0$ for each i.

(iv) If X is a linear lattice, and each x_i is non-zero, then $\{x_i\}$ is linearly independent.

<u>Proof</u>. (i) Let $y = x_1^+ + \ldots + x_n^+$, $z = x_1^- + \ldots + x_n^-$. Now $x_i^+ \wedge x_j^- = 0$ for all i,j, so by 2.4.1, $x_i^+ \wedge z = 0$. Therefore, by 2.4.1 again, $y \wedge z = 0$. But $y - z = x$, so, by 2.3.1(i), $y = x^+$ and $z = x^-$.

(ii) The first equality is obtained by adding the two equalities in (i). By 2.2.3, $|x_1| + |x_2| = |x_1| \vee |x_2|$. Suppose that, for some $k < n$,

$$|x_1| + \ldots + |x_k| = |x_1| \vee \ldots \vee |x_k|.$$

Denote this element by y. Then $y \wedge |x_{k+1}| = 0$, by 2.4.1, so $y \vee |x_{k+1}| = y + |x_{k+1}|$. By induction, it follows that

$$|x_1| + \ldots + |x_n| = |x_1| \vee \ldots \vee |x_n|.$$

(iii) If $x_i \notin P$ for some i, then $x_i^- > 0$, so $x^- > 0$, and $x \notin P$.

(iv) Suppose that $\sum_{i=1}^{n} \lambda_i x_i = 0$. The elements $\lambda_i x_i$ are pairwise disjoint, so by (iii), $\lambda_i = 0$ for each i.

Let X be a set with an antisymmetric ordering. We say that a subset E of X <u>admits suprema</u> if for each subset A of E having a supremum in X, it is true that sup $A \in E$. If E is order-convex in X, it is clearly enough to know that, for each such A, some upper bound of A is in E. As an immediate corollary of 2.2.12, we have:

<u>2.4.3</u>. The following condition is sufficient for a lattice subgroup E of a commutative lattice group X to admit suprema: if A is a subset of $E \cap P$ that admits $\vee$ and has a supremum in X, then sup $A \in E$.

A <u>band</u> (the Russian term is <u>component</u>) in a commutative lattice group is defined to be a solid subgroup that admits suprema. We recall that a solid subgroup of a linear lattice is necessarily a linear subspace.

<u>Example</u>. Consider m with the usual order. Let $E = \{x: \xi_1 = 0\}$. Then E is a band in m. On the other hand, c_o (which is a solid subspace) does not admit suprema, since, with our usual notation, $e_n \in c_o$, $e = \sup\{e_n\}$, and $e \notin c_o$.

If A is a subset of a commutative lattice group X, then the set of elements of X disjoint to every element of A will be called the <u>annihilator</u> of A, and denoted by $A^{\perp}$.

<u>2.4.4</u>. If A is a subset of a commutative lattice group, then $A^{\perp}$ is a band.

<u>Proof</u>. It is obvious that $A^{\perp}$ is solid. It follows from 2.4.1 that $A^{\perp}$ is a subgroup. Suppose that $B \subseteq A^{\perp} \cap P$ and that B has a supremum b_0 in X. By 2.2.7, for a in A,

$$|a| \wedge b_0 = \sup \{ |a| \wedge b : b \in B \}$$
$$= 0,$$

so $b_0 \in A^{\perp}$. By 2.4.3, it follows that $A^{\perp}$ admits suprema.

<u>2.4.5. COROLLARY</u>. If E is the smallest band containing A, then $E^{\perp} = A^{\perp}$.

<u>Proof</u>. Clearly, $E^{\perp} \subseteq A^{\perp}$. If $x \in A^{\perp}$, then $A \subseteq \{x\}^{\perp}$, which is a band, so $E \subseteq \{x\}^{\perp}$. Hence $x \in E^{\perp}$.

<u>2.4.6</u>. If X is a commutative lattice group, and E_1, E_2 are subgroups whose ordered direct sum is X, then E_1 and E_2 are bands.

<u>Proof</u>. By 1.4.2, E_1 and E_2 are solid. Hence if $x_i \in E_i$ $(i = 1,2)$, then $|x_1| \wedge |x_2| \in E_1 \cap E_2$, so $|x_1| \wedge |x_2| = 0$. Therefore $E_1 \subseteq E_2^{\perp}$. Since $E_2 \cap E_2^{\perp} = \{0\}$, it follows that $E_1 = E_2^{\perp}$. Similarly, $E_2 = E_1^{\perp}$.

<u>2.4.7</u>. Let X be an Archimedean commutative lattice group. For any subset A of X, $A^{\perp\perp}$ is the band generated by A.

<u>Proof</u>. Let E be the smallest solid subgroup containing A. Then $E^{\perp} = A^{\perp}$, by 2.4.5, and the band generated by A is the band generated by E. Denote this band by F. Suppose that there is an element u in $E^{\perp\perp} \sim F$. Let $B = \{ v \in E : 0 \leq v \leq u \}$. If B has a supremum, then the supremum is in F, so u is not the supremum of B. Hence there is an

element w' such that $B \leq w'$ and $u \leq w'$. Let $w = u \wedge w'$. Then $B \leq w \leq u$. Hence $0 < u - w < u$, so $u - w \in E^{\perp\perp}$. Therefore $u - w \notin E^{\perp}$, so there exists y in E such that $y \wedge (u - w) > 0$. Let $x = y \wedge (u - w)$. Then $x \in E$, and $0 < x \leq u - w$. We prove by induction that $nx \leq u$ $(n \in \omega)$, showing that X is not Archimedean. Suppose that $(n - 1)x \leq u$. Then $(n - 1)x \in B$, so $(n - 1)x \leq w$. Thus $nx \leq w + (u - w) = u$, as required.

In particular, if E is a band, then $E^{\perp\perp} = E$.

Example. Consider R^2 with the lexicographic ordering. Let $E = \{(0,\eta) : \eta \in R\}$. Then E is a band, and $E^{\perp} = \{0\}$, so $E^{\perp\perp} = R^2$.

Actually, the converse of 2.4.7 is true: if $E^{\perp\perp} = E$ for every band E, then the ordering is Archimedean (see Luxemburg - Zaanen (1), sect. 9).

Freudenthal units

An element e of a commutative lattice group is said to be a Freudenthal unit if $\{e\}^{\perp} = \{0\}$ (equivalently, if $e \wedge x = 0$ implies $x = 0$).

2.4.8. Any order-unit is a Freudenthal unit.

Proof. Suppose that e is an order-unit, and that $e \wedge x = 0$. There exists a positive integer n such that $x \leq ne$, or $(ne) \wedge x = x$. But $(ne) \wedge x = 0$, by 2.4.1.

Examples. In the sequence spaces s, m, c_o, any sequence $\{\xi_n\}$ having $\xi_n > 0$ for each n is a Freudenthal unit. (We recall that s and c_o have no order-units.)

2.5. Order-completeness

Order-completeness was defined in section 0.1. A commutative lattice group is said to be order-σ-complete if each countable majorised subset has a supremum (equivalently, if each countable minorised subset has an infimum). This clearly forms another intermediate stage in the scale of interpolation properties.

The following terminology is also in use: Dedekind complete or σ-complete by Luxemburg and Zaanen; K-space or K_σ-space by Russian writers; completely continuous or continuous by Japanese writers. There is an extensive Russian and Japanese literature on such spaces. The treatment given here is confined to elementary topics.

From 2.2.12, we have at once:

2.5.1. The following condition is sufficient for a commutative lattice group to be order complete; every majorised set of positive elements that admits $\vee$ has a supremum.

Order-σ-completeness is also equivalent to a weaker statement:

2.5.2. The following condition is sufficient for a commutative lattice group to be order-σ-complete: every majorised increasing sequence of positive elements has a supremum.

Proof. Suppose that the condition holds, and that $x_n \leq a$ $(n \in \omega)$. Let $x'_n = x_n - x_1$ $(n \in \omega)$, and let $y_n = x'_1 \vee \ldots \vee x'_n$. Then $0 \leq y_n \leq y_{n+1} \leq a - x_1$ $(n \in \omega)$, so, by hypothesis, $\{y_n\}$ has a supremum. Since $\{y_n\}$ has the same upper bounds as $\{x'_n\}$, the result follows.

2.5.3. An order-σ-complete commutative lattice group is Archimedean.

Proof. Suppose that $nx \leq y$ $(n \in \omega)$. Let $z = \sup\{nx : n \in \omega\}$. Then $x + z = \sup\{nx : n = 2,3, \ldots\} \leq z$, so $x \leq 0$.

Examples

(i) The space of all real-valued functions on a set S is order-complete. If E is bounded above, then sup E = y, where $y(s) = \sup \{x(s) : x \in E\}$ $(s \in S)$.

(ii) In particular, the space s of all real sequences (with its usual ordering) is order-complete, with a pointwise definition of suprema. The same is true of its order-convex subspaces, for instance m, c_o, l_1, F.

(iii) The lexicographic ordering of R^2 is not order-σ-complete, since it is not Archimedean.

(iv) The space C[0,1] is not order-complete. For instance, if x_n is the piecewise linear function taking the value 1 on $[0, \frac{1}{2} - \frac{1}{n}]$ and the value 0 on $[\frac{1}{2}, 1]$, then $\{x_n\}$ has no supremum. At the end of this section, we find necessary and sufficient conditions for a space of continuous real-valued functions to be order-complete.

An example of an order-σ-complete space that is not order-complete is given after 2.5.6.

Decomposition into bands

Order-complete spaces furnish us with the long-awaited converse to 1.4.2:

2.5.4. THEOREM (Riesz). If X is an order-complete commutative lattice group, and E is a band in X, then X is the ordered direct sum of E and $E^\perp$. If $x \geq 0$, and

$$y = \sup \{u \in E : 0 \leq u \leq x\}$$

then $y \in E$, $x - y \in E^\perp$.

Proof. It is clear that $0 \leq y \leq x$, and $y \in E$, since E is a band. Take a in E, and let $b = |a| \wedge (x - y)$. Then $0 \leq b \leq |a|$, so $b \in E$. Hence $b + y \in E$. Also, $0 \leq b + y \leq x$, so $b + y \leq y$, by definition. Hence $b = 0$, so $x - y \in E^{\perp}$, as stated.

2.5.5. COROLLARY. If A is any subset of an order-complete commutative lattice group X, then $A^{\perp} + A^{\perp\perp} = X$.

For bands generated by one element, order-σ-completeness is sufficient:

2.5.6. Let X be an order-σ-complete commutative lattice group, and let u, x be positive elements of X. Let $y = \sup\{x \wedge nu : n \in \omega\}$. Then $y \in \{u\}^{\perp\perp}$, $x - y \in \{u\}^{\perp}$.

Proof. Since $\{u\}^{\perp\perp}$ is a band, it contains y. By 2.2.7,

$$y = y \wedge y = \sup\{x \wedge y \wedge nu : n \in \omega\}$$
$$= \sup\{y \wedge nu : n \in \omega\}, \text{ since } y \leq x.$$

Let $z = (x - y) \wedge u$. Then

$$z + (y \wedge nu) = (z + y) \wedge (z + nu)$$
$$\leq x \wedge (n + 1)u.$$

Taking suprema on both sides, we see that $z + y \leq y$. Hence $z = 0$, as required.

In particular, if x is a positive element of $\{u\}^{\perp\perp}$, then $x = \sup\{x \wedge nu : n \in \omega\}$. If u is a Freudenthal unit, this equality holds for each positive element x.

Let us say, temporarily, that a space has the decomposition property if, for each band E, we have $E + E^{\perp} = X$. We now give an example of (1) a space that is order-σ-complete but does not have the decomposition property, (2) a space that has the decomposition property but is not order-σ-complete. By 2.5.4, the space in example (1) cannot be order-complete. The two examples (both due to Luxemburg and Zaanen) are surprisingly similar.

(1) Let X be the set of bounded real functions x on $[0,1]$ for which $\{s : x(s) \neq x(0)\}$ is at most countable. The

natural ordering makes X a linear lattice. If $\{x_n\}$ is a majorized sequence in X, then

$\{s : x_n(s) \neq x_n(0) \text{ for some } n \text{ in } \omega\}$

is at most countable, so the pointwise supremum of $\{x_n\}$ is in X. Hence X is order-σ-complete. Let A be the set of elements of X that vanish on $[0,\frac{1}{2}]$. It is easily verified that A is a band, and that $A^{\perp}$ is the set of elements of X that vanish on $(\frac{1}{2},1]$. For x in $A^{\perp}$, $x(0) = 0$. Hence this is also true for x in $A + A^{\perp}$, showing that $A + A^{\perp} \subset X$.

(2) Let X be the set of real sequences that take only a finite number of different values. The natural ordering makes X a linear lattice. Let

$$x_n = (1,\tfrac{1}{2}, \ldots, \tfrac{1}{n},0,\ldots).$$

Then $\{x_n\}$ is majorised, but has no supremum. Let A be a band in X, and let

$$N_1 = \{n \in \omega : \xi_n \neq 0 \text{ for some x in } A\},$$

$$N_2 = \omega \sim N_1.$$

Then $A^{\perp}$ is the set of sequences in X that vanish on N_1. By 2.4.7, $A = A^{\perp\perp}$, so A is the set of sequences in X that vanish on N_2. It now follows easily that $A + A^{\perp} = X$.

Linear mappings into order-complete spaces

If X is a linear space and Y is an ordered linear space, then a mapping p from X to Y is said to be <u>sublinear</u> if $p(\lambda x) = \lambda p(x)$ $(x \in X, \lambda \geq 0)$, and $p(x + y) \leq p(x) + p(y)$ $(x,y \in X)$. By a slight adaptation of the standard proof of the Hahn-Banach theorem (see, e.g., Day (1), p.9), we can prove:

<u>2.5.7</u>. Let X be a linear space, Y an order-complete linear lattice,

and p a sublinear mapping from X to Y. Suppose that E is a linear subspace of X, and that f is a linear mapping from E to Y satisfying $f(x) \leq p(x)$ $(x \in E)$. Then there is a linear mapping $\bar{f}$ from X to Y that extends f and satisfies $\bar{f}(x) \leq p(x)$ $(x \in X)$.

In the usual way, we can deduce:

<u>2.5.8. COROLLARY</u>. Let X be a linear space, Y an order-complete linear lattice, and p a sublinear mapping from X to Y. Then, given a in X, there exists a linear mapping f from X to Y such that $f(x) \leq p(x)$ $(x \in X)$ and $f(a) = p(a)$.

Either of the two proofs given for the monotone extension theorem 1.6.1 can be adapted to give:

<u>2.5.9</u>. Suppose that X is an ordered linear space and Y an order-complete linear lattice. Let E be a cofinal linear subspace of X, and f a monotonic linear mapping from E to Y. Then there is a monotonic linear mapping from X to Y that extends f.

We can deduce the following monotone extension theorem, in which the domain, not the range, is required to be order-complete:

<u>2.5.10</u>. Suppose that X is a linear lattice and Y an ordered linear space. Let E be an order-complete, cofinal linear subspace of X. Then a monotonic linear mapping f from E to Y has a monotonic linear extension to X.

<u>Proof</u>. By 2.5.9, the identity mapping in E can be extended to a monotonic linear mapping p from X to E. Let $f(x) = f(p(x))$ $(x \in X)$. Then f has the required properties.

Other results on linear functionals, for instance 1.5.8, can be generalised in the same way.

Conversely, it can be shown that if Y is such that the conclusion of 2.5.7 or 2.5.9 holds for all X and E, then Y is order-complete (see Day (1), p. 105-106, Bonnice - Silverman (1)). Antisymmetry of the order in Y is not needed in these results.

Spaces of continuous functions

Let S be a topological space. We denote by C(S) the set of continuous real-valued functions on S, and by $C_B(S)$ the set of bounded functions in C(S). Since $C_B(S)$ is an order-convex subspace of C(S), it is order-complete whenever C(S) is.

A topological space is said to be extremally disconnected if the closure of every open set is open.

2.5.11. THEOREM (Stone). If S is an extremally disconnected topological space, then C(S) is order-complete.

Proof. It is sufficient to show that every family F of non-negative functions in C(S) has an infimum. For each $\lambda > 0$, let

$$G_\lambda = \{s : f(s) < \lambda \text{ for some } f \text{ in } F\}.$$

Then (i) G_λ is open, (ii) $G_\lambda \subseteq G_\mu$ whenever $\lambda \leq \mu$, and (iii) $\bigcup\{G_\lambda : \lambda > 0\} = S$. For each s in S, we can therefore define

$$g(s) = \inf \{\lambda > 0 : s \in \overline{G}_\lambda\}.$$

If $\mu > 0$, then

$$\{s : g(s) < \mu\} = \bigcup\{\overline{G}_\lambda : \lambda < \mu\},$$

which is open, since S is extremally disconnected. If $\mu \geq 0$, then

$$\{s : g(s) \leq \mu\} = \bigcap\{\overline{G}_\lambda : \lambda < \mu\},$$

which is closed. It follows that g is continuous.

If $\lambda < g(s)$, then $s \notin G_\lambda$, so $f(s) \geq \lambda$ $(f \in F)$. Hence $g \leq F$. Suppose that $h \in C(S)$ and $h \leq F$. If $s \in G_\lambda$, then $h(s) < \lambda$. Since h is continuous, it follows that $\overline{G}_\lambda \subseteq \{s : h(s) \leq \lambda\}$. If $g(s) < \lambda$, then $s \in \overline{G}_\lambda$, so $h(s) \leq \lambda$. Hence $h \leq g$, so g is the infimum of F in

C(S).

It is easily verified that, with the notation used in the theorem,

$$g(s) = \sup \{\inf F(N) : N \text{ a neighbourhood of } s\},$$

where

$$F(N) = \{f(t) : f \in F, t \in N\}.$$

In completely regular spaces, the converse holds:

2.5.12. If S is a completely regular space, then the following statements are equivalent:

(i) S is extremally disconnected.

(ii) C(S) is order-complete.

(iii) $C_B(S)$ is order-complete.

Proof. (i) implies (ii), by 2.5.11, and (ii) implies (iii), since $C_B(S)$ is order-convex in C(S). To prove that (iii) implies (i), take an open proper subset G of S. Let F be the set of all continuous functions on S with values in [0,1] that take the value 0 off G. Then F is majorised by the unit function, so has a supremum in $C_B(S)$, say g. If $s \in G$, then, by complete regularity, there exists f in F with f(s) = 1. Hence g(s) = 1, and since g is continuous, it follows that g takes the value 1 on $\overline{G}$. If $t \notin \overline{G}$, then there is a continuous function h on S with values in [0,1] such that h(t) = 0 and h = 1 on G. Then $h \geq F$, so $h \geq g$. Hence g(t) = 0. Thus g is the characteristic function of $\overline{G}$. Since g is continuous, it follows that $\overline{G}$ is open.

2.6. Linear mappings and functionals

The first two theorems of this section are better known as theorems on linear functionals, but are given here in the form applicable to linear mappings with values in an order-complete linear lattice. This complicates the statements slightly, but has virtually no effect on the proofs.

The Riesz decomposition theorem

2.6.1. THEOREM (Riesz). Let X be a Riesz space with positive wedge P, and let Y be an order-complete linear lattice. Let $(X,Y)^b$ be the set of order-bounded linear mappings from X to Y. Then $(X,Y)^b$ is an order-complete linear lattice, and the following statements hold:

(i) For x in P and f,g in $(X,Y)^b$,

$f^+(x) = \sup f[0,x]$;

$f^-(x) = \sup f[-x,0]$;

$|f|(x) = \sup f[-x,x]$;

$(f \vee g)(x) = \sup \{f(u) + g(x-u) : 0 \leq u \leq x\}$.

(ii) If F is a majorised, upward-directed subset of $(X,Y)^b$, then $\sup F = g$, where $g(x) = \sup \{f(x) : f \in F\}$.

Proof. (i) Take f in $(X,Y)^b$, and write $\sup f[0,x] = g(x)$ $(x \in P)$. Then $g(\lambda x) = \lambda g(x)$ $(\lambda > 0,\ x \in P)$, and if $x,y \in P$, then $[0,x] + [0,y] = [0,x+y]$, by 2.1.3, so $g(x+y) = g(x) + g(y)$, by 1.1.7. By 1.5.6, g can be extended to a linear mapping on X by defining $g(u-v) = g(u) - g(v)$ $(u,v \in P)$. Clearly, $g \geq 0$ and $g \geq f$. If $h \geq 0$ and $h \geq f$, then $h(x) \geq h(y)$ for all y in $[0,x]$, so $h(x) \geq g(x)$ $(x \in P)$. Hence the ordering of $(X,Y)^b$ is a lattice ordering, and $g = f^+$. Also, we have:

$$\begin{aligned} f^-(x) &= \sup \{-f(u) : 0 \leq u \leq x\} \\ &= \sup \{f(-u) : -x \leq -u \leq 0\}; \\ |f|(x) &= 2f^+(x) - f(x) \\ &= \sup f[0, 2x] - f(x) \\ &= \sup f[-x, x]; \\ (f \vee g)(x) &= (f-g)^+(x) + g(x) \\ &= g(x) + \sup \{f(u) - g(u) : 0 \leq u \leq x\} \\ &= \sup \{f(u) + g(x-u) : 0 \leq u \leq x\}. \end{aligned}$$

(ii) With F as stated, let $g(x) = \sup\{f(x) : f \in F\}$. Then g is clearly positive homogeneous. We show that it is also additive, so that it can be extended to a linear mapping on X, by 1.5.6, giving the result. Take x,y in P. Then

$$g(x + y) = \sup\{f(x) + f(y) : f \in F\}$$
$$= \sup\{f_1(x) + f_2(y) : f_1, f_2 \in F\},$$
since F is directed upwards
$$= g(x) + g(y), \text{ by 1.1.7.}$$

In particular, every order-bounded linear functional on a Riesz space X is the difference between two monotonic linear functionals, (cf. section 1.5, ex.(v)), and the space X^b of order-bounded linear functionals is an order-complete linear lattice. When X is a lattice, we have:

<u>2.6.2. COROLLARY</u>. If X is a linear lattice and A is a solid subset of X, then the polar of A in X^b is solid.

<u>Proof</u>. Take f in A^o and a in A. Then $[-|a|, |a|] \subseteq A$, so $|f|(|a|) = \sup f\,[-|a|, |a|] \leqslant 1$. If $|g| \leqslant |f|$, then $g(a) \leqslant |g|(|a|)$, by 1.5.4, so $g(a) \leqslant |f|(|a|) \leqslant 1$. Hence $g \in A^o$.

<u>An extension theorem</u>

Let X be a linear lattice, Y an ordered linear space. A sublinear mapping p from X to Y will be said to be <u>lattice-monotonic</u> if $|x_1| \leqslant |x_2|$ implies $p(x_1) \leqslant p(x_2)$. If the ordering in Y is anti-symmetric, it is clear that the following conditions are necessary and sufficient for p to be lattice-monotonic:

(i) $0 \leqslant x_1 \leqslant x_2$ implies $p(x_1) \leqslant p(x_2)$ and

(ii) $p(|x|) = p(x)$ for all x in X.

From (ii) we see that $p(\lambda x) = |\lambda| p(x)$ for all λ in R. A lattice-monotonic mapping into R is called a <u>lattice seminorm</u>.

The following theorem, due to Luxemburg and Zaanen, may be regarded as the variant of the Hahn-Banach theorem suitable to linear lattices.

<u>2.6.3. THEOREM</u>. Let X be a linear lattice, Y an order-complete linear lattice, and E a linear sublattice of X. Suppose that p is a lattice-monotonic sublinear mapping from X to Y, and that f is a monotonic linear mapping from E to Y such that $f(x) \leq p(x) \quad (x \in E)$. Then there is a monotonic linear mapping $\bar{f}$ from X to Y such that $\bar{f}(x) = f(x) \quad (x \in E)$ and $\bar{f}(x) \leq p(x) \quad (x \in X)$.

<u>Proof</u>. Let $q(x) = p(x^+) \quad (x \in X)$. Then $q(\lambda x) = \lambda q(x)$ for $\lambda \geq 0$, and since $(x_1 + x_2)^+ \leq x_1^+ + x_2^+$, we have

$$q(x_1 + x_2) \leq p(x_1^+ + x_2^+) \leq p(x_1^+) + p(x_2^+) = q(x_1) + q(x_2).$$

For x in E, $f(x) \leq f(x^+) \leq p(x^+) = q(x)$. Hence, by 2.5.7, there is a linear extension $\bar{f}$ of f defined on X and satisfying $\bar{f}(x) \leq q(x)$ there. Since p is lattice-monotonic and $x^+ \leq |x|$, we have $q(x) \leq p(x) \quad (x \in X)$. For $x \leq 0$, $q(x) = 0$, so $\bar{f}(x) \leq 0$.

It is instructive to compare this theorem with 2.5.9. Here, the subspace E is not required to be cofinal, and we start by assuming the existence of the mapping p. If f is a monotonic linear mapping, and $r(x) = f(|x|) \ (x \in X)$, then r is lattice-monotonic. Hence the existence of non-zero monotonic linear mappings is equivalent to the existence of non-zero lattice-monotonic sublinear mappings.

<u>An existence theorem</u>

<u>2.6.4. THEOREM</u>. Let X,Y be linear lattices, and suppose that either X or Y is order-σ-complete. Let f be a monotonic linear mapping from X to Y. Then, given a positive element a of X, there exists a linear mapping g from X to Y such that $0 \leq g \leq f$, $g(a) = f(a)$ and $g(x) = 0$ for all x in $\{a\}^\perp$.

Proof. (i) Suppose that X is order-σ-complete. Let p be the projection onto $\{a\}^{\perp\perp}$ along $\{a\}^{\perp}$ (as in 2.5.6). Then $p(a) = a$, and if $x \geq 0$, then $0 \leq p(x) \leq x$. Define $g(x) = f(p(x))$ $(x \in X)$. Then g has the required properties.

(ii) Suppose that Y is order-σ-complete. Let P denote the positive cone in X. For x in P, we can define

$$g(x) = \sup \{f(x \wedge na) : n \in \omega\}.$$

Then $0 \leq g(x) \leq f(x)$ $(x \in P)$, $g(a) = f(a)$, and if $x \wedge a = 0$, then $g(x) = 0$. We show that g is additive and positive homogeneous on P; 1.5.6. then shows that g has a linear extension to X with the required properties. Positive homogeneity is straightforward. Now

$$(x \wedge ma) + (y \wedge na) \leq (x + y) \wedge (m + n)a,$$

so $g(x) + g(y) \leq g(x + y)$. By 2.2.5,

$$(x + y) \wedge (na) \leq (x \wedge na) + (y \wedge na),$$

so we have the reverse inequality as well.

We can deduce some equalities dual to those in 2.6.1(i):

2.6.5. COROLLARY. Let X,Y be linear lattices, one of which is order-σ-complete, and let f be a monotonic linear mapping from X to Y. Then, for a in X, we have:

$$f(a^+) = \sup \{g(a) : 0 \leq g \leq f\},$$

$$f(a^-) = \sup \{g(a) : -f \leq g \leq 0\},$$

$$f(|a|) = \sup \{g(a) : -f \leq g \leq f\}.$$

Proof. If $0 \leq g \leq f$, then $g(a) \leq g(a^+) \leq f(a^+)$. By 2.6.4, there exists g such that $0 \leq g \leq f$, $g(a^+) = f(a^+)$ and $g(a^-) = 0$. Then $g(a) = f(a^+)$.

The second equality follows at once. To prove the third one, note that $|a| = 2a^+ - a$. Thus

$$f(|a|) = \sup \{ h(a) : 0 \leqslant h \leqslant 2f\} - f(a)$$

$$= \sup \{ g(a) : -f \leqslant g \leqslant f\}.$$

In the case when Y is order-complete, (e.g. when considering linear functionals), a very short proof of these equalities can be given by considering the sublinear mappings q,r, where $q(x) = f(x^+)$, $r(x) = f(|x|)$ $(x \in X)$, and applying 2.5.8.

<u>Lattice homomorphisms</u>

A mapping from one lattice to another is called a <u>lattice homomorphism</u> if it preserves the lattice operations. Lattice homomorphisms are monotonic, since $x \leqslant y$ implies that $x \vee y = y$, and hence that $f(x) \vee f(y) = f(y)$. If a lattice homomorphism has an inverse, then this is also a lattice homomorphism (and so, in particular, is monotonic).

The relation $x \vee y + x \wedge y = x + y$ shows that it is sufficient for a linear mapping f to be a lattice homomorphism if it preserves either $\vee$ or $\wedge$. Since $x \vee y = (x - y)^+ + y$, it is sufficient if $f(x^+) = (f(x))^+$ for all x.

<u>2.6.6</u>. If X,Y are linear lattices, and f is a linear mapping from X to Y, then the following statements are equivalent:

(i) f is a lattice homomorphism;

(ii) $f(x^+) \wedge f(x^-) = 0$ for all x in X;

(iii) if $x \wedge y = 0$, then $f(x) \wedge f(y) = 0$;

(iv) $f(|x|) = |f(x)|$ for all x in X.

<u>Proof</u>. (i) => (ii). Immediate, since $x^+ \wedge x^- = 0$.

(ii) =>(iii). If $x \wedge y = 0$, then, by 2.3.3, $(x - y)^+ = x$, $(x - y)^- = y$.

(iii) => (i). For any x,y in X, we have

$(x - x\wedge y)\wedge(y - x\wedge y) = x\wedge y - x\wedge y = 0.$

Thus (iii) implies that $f(x)\wedge f(y) = f(x\wedge y)$.

(i) => (iv). For any x, $f(|x|) = f(x)\vee f(-x) = |f(x)|$.

(iv) => (i). If (iv) holds, then, for any x in X,

$$2f(x^+) = f(|x| + x) = |f(x)|+f(x) = 2(f(x))^+.$$

2.6.7. If X,Y are linear lattices, with positive cones P,Q respectively, and f is a linear mapping on X onto Y, then the following statements are equivalent:

(i) f is a lattice homomorphism;

(ii) the kernel of f is solid and $f(P) = Q$.

Proof. (i) => (ii). If $f(u) = 0$ and $|v| \leq |u|$, then

$|f(v)| = f(|v|) \leq f(|u|) = |f(u)| = 0.$

Take y in Q. There exists x in X such that $f(x) = y$. Then $f(x^+) = y^+ = y$, so $y \in f(P)$.

(ii) => (iii). Take u,v in X such that $u \wedge v = 0$. There exists w in P such that $f(u) \wedge f(v) = f(w)$. Then $f(u-w)$ and $f(v-w)$ are in Q, so there exist a,b such that $f(a) = f(b) = 0$ and $u - w \geq a$, $v - w \geq b$. Then $a \wedge b \leq -w$. Since the kernel of f is solid, $f(a \wedge b) = 0$, so $f(w) \leq 0$. Hence $f(w) = 0$, and the result follows, by 2.6.6(iii).

In particular, if f is a non-zero linear functional with solid kernel, then, by 1.5.5, either f or $-f$ is monotonic, so is a lattice homomorphism.

We give two more results which are very easy to prove using 2.6.6(iii).

2.6.8. If X is a linear lattice, and E_1, E_2 are bands whose direct sum is X, then the projection onto E_1 along E_2 is a lattice homomorphism.

Proof. Denote the projection by p, and take x,y in X such that $x \wedge y = 0$. Since $0 \leq p(x) \leq x$ and $0 \leq p(y) \leq y$, we have $p(x) \wedge p(y) = 0$.

2.6.9. Let X,Y be linear lattices, and let f,g be linear mappings from X to Y such that $0 \leq g \leq f$. If f is a lattice homomorphism, then so is g.

Proof. Take u,v in X such that $u \wedge v = 0$. Then

$$0 \leq g(u) \wedge g(v) \leq f(u) \wedge f(v) = 0.$$

Example.

Let X be a linear lattice of real-valued functions on a set S. For each s in S, a lattice homomorphism f_s from X to R is defined by: $f_s(x) = x(s) \quad (x \in X)$.

A lattice homomorphism does not necessarily preserve infima of countable sets (though it is elementary that a lattice isomorphism between two lattices preserves all suprema and infima). Let x_n be the element of C[0,1] defined by $x_n(s) = (1 - ns) \vee 0 \; (0 \leq s \leq 1)$. Then $\inf \{x_n\} = 0$, but (with the above notation), $f_0(x_n) = 1$ for each n.

Embedding in the second conjugate

2.6.10. Let X be a linear lattice, and let Y be a solid subspace of X^b that separates points of X. Then the natural embedding map of X into Y' is a linear lattice homomorphism.

Proof. The embedding map is i, where $i(x)(f) = f(x) \quad (f \in Y)$. We show that, for positive f in Y, $(i(x))^+(f) = i(x^+)(f)$, from which the result follows. By 2.6.1,

$$(i(x))^+(f) = \sup \{i(x)(g) : g \in Y \text{ and } 0 \leq g \leq f\}.$$

Since Y is solid in X^b, this is equal to

$$\sup \{i(x)(g) : g \in X^b \text{ and } 0 \leq g \leq f\}$$
$$= \sup \{g(x) : g \in X^b \text{ and } 0 \leq g \leq f\}.$$

By 2.6.5, this is equal to $f(x^+)$, as required.

2.7. Discrete and extremal monotonic linear mappings

It follows from 1.8.1 that a monotonic linear functional on a linear lattice is extremal iff it is a lattice homomorphism (a more direct proof can be given on the lines of the remark following 2.6.5). To obtain a corresponding result for linear mappings, we make the following definition. Let X be a linear lattice, Y a linear space. A linear mapping f from X to Y is said to be <u>discrete</u> if, for each x in X, either $f(x^+) = 0$ or $f(x^-) = 0$.

If Y is also a linear lattice, then 2.6.6 shows that a discrete monotonic linear mapping from X to Y is a lattice homomorphism. The converse is true for linear functionals, but not for linear mappings in general, as is shown by the identity mapping.

<u>2.7.1</u>. The following condition is necessary and sufficient for a linear mapping f on a linear lattice to be discrete: for each pair of elements x,y, one of the two elements $f(x \vee y)$, $f(x \wedge y)$ is equal to $f(x)$, and the other is equal to $f(y)$.

<u>Proof</u>. Sufficiency is obvious. Necessity follows from 2.3.3.

<u>2.7.2</u>. Let X,Y be linear lattices, and suppose that either X or Y is order-σ-complete. Then an extremal monotonic linear mapping from X to Y is discrete.

<u>Proof</u>. Suppose that f is an extremal monotonic linear mapping from X to Y, and that $f(a^+) \neq 0$. By 2.6.4, there exists a linear mapping g from X to Y such that $0 \leq g \leq f$, $g(a^+) = f(a^+)$ and

$g(a^-) = 0$. Since f is extremal, we must have $g = f$. Hence $f(a^-) = 0$.

We now use 1.5.11 to show that, under quite general conditions, discrete monotonic linear mappings have one-dimensional ranges (so that they are essentially nothing more than linear functionals).

2.7.3. Let X be a linear lattice, and Y a linear space with an almost Archimedean ordering. If f is a discrete monotonic linear mapping from X to Y, then there exist a discrete linear functional h on X and a positive element b of Y such that $f(x) = h(x)b \quad (x \in X)$.

Proof. The range of f is almost Archimedean, since it is a subspace of Y. If $f(x^+) = 0$, then $f(x) = -f(x^-) \leq 0$. If $f(x^-) = 0$, then $f(x) = f(x^+) \geq 0$. Hence the range of f is totally ordered, so, by 1.5.11, its dimension is one or zero. The result follows.

We notice that every linear mapping on a totally ordered linear space is discrete, since for each x, either $x^+ = 0$ or $x^- = 0$. This shows that 2.7.3 can fail if Y is not almost Archimedean (consider R^2 with the lexicographic ordering).

Combining the two preceding results, we have:

2.7.4. THEOREM. Let X be a linear lattice, Y an order-σ-complete linear lattice, and f a linear mapping from X to Y. Let P,Q denote the positive cones in X,Y. Then the following statements are equivalent:

(i) f is an extremal monotonic linear mapping from X to Y;

(ii) there exist extremal elements h of P^o and b of Q such that $f(x) = h(x)b \quad (x \in X)$.

Proof. (i) => (ii). By 2.7.2, f is discrete. By 2.5.3, Y is Archimedean, so 2.7.3 applies to show that there exist h in P^o and b in

Q such that $f(x) = h(x)b$ $(x \in X)$. Take h' such that $0 \leq h' \leq h$, and define $f'(x) = h'(x)b$ $(x \in X)$. Then $0 \leq f' \leq f$, so there exists λ in R such that $f' = \lambda f$. Hence $h' = \lambda h$, and h is an extremal element of P^o. Similarly b is an extremal element of Q.

(ii) => (i). With f as described, suppose that $0 \leq g \leq f$. For each x in P, $0 \leq g(x) \leq h(x)b$, so there is a unique real number k(x) such that $g(x) = k(x)b$. Since $P - P = X$, the functional k can be extended so that this equality holds for all x in X. Then $0 \leq k \leq h$, so $k = \lambda h$ for some λ in R. Then we have $g = \lambda f$.

Clearly, 2.7.4 also holds under the assumption that X is order-σ-complete and Y is Archimedean. In fact, it also holds if X is any ordered linear space and Y is order-complete (see Jameson (4)).

Finally, the following result is of interest:

__2.7.5.__ If f is a discrete, order-bounded linear functional on a linear lattice, then either f or -f is monotonic.

__Proof.__ Suppose that there exist positive elements y', z' such that $f(y') = 1$, $f(z') = -1$. Let $x = y' + z'$, and let $\sup f[0,x] = \alpha$. Then $\alpha \geq 1$, and $\inf f[0,x] = -\alpha$, by 1.5.8. Take y,z in $[0,x]$ such that $f(y) > \frac{2}{3}\alpha$, $f(z) < -\frac{2}{3}\alpha$. Let $u = y \wedge z$. By 2.7.1, $f(u) = f(y)$ or $f(u) = f(z)$. If $f(u) = f(z)$, then $f(y-u) = f(y) - f(z) > \frac{4}{3}\alpha$, and if $f(u) = f(y)$, then $f(z - u) = f(z) - f(y) < -\frac{4}{3}\alpha$. Since $y - u$ and $z - u$ are in $[0,x]$, this is a contradiction.

CHAPTER 3

ORDERED TOPOLOGICAL LINEAR SPACES

Given an ordered linear space, it is possible to consider topologies induced in some way by the ordering (see sections 3.7, 3.9, 3.10), but on the whole it is more interesting to consider linear spaces in which an ordering and a topology are given separately. Such structures turn out to be rich in duality theory, motivated by the consideration that the larger P is, the smaller P^o becomes, and conversely. The main conditions linking the topology and the ordering which will occupy us in this chapter are:

(1) X is locally order-convex;

(1') P has a bounded base B such that $0 \notin \overline{B}$;

(2) for each U in Ⓝ(X), $P \cap U - P \cap U$ is in Ⓝ(X);

(2') P has an interior point.

Among these conditions, (1') implies (1) and (2') implies (2). Properties (1) and (1') are inherited by any wedge contained in P, while properties (2) and (2') are inherited by any wedge containing P. Properties (1) and (2) can be regarded, respectively, as topological embellishments of the conditions that the ordering should be anti-symmetric and directed. Either (1) combined with (2'), or (2) combined with (1'), implies that the topology is normable.

For normed spaces, the duality situation is very satisfactory (the section on normed spaces, 3.6, can be read in isolation by readers unfamiliar with topological linear spaces). We find that P satisfies (1) iff P^o satisfies (2), and when X is complete and P closed, that P satisfies (2) iff P^o satisfies (1). The more extreme conditions (1') and (2') are similarly related, subject to slight restrictions (section 3.8). The position in topological linear spaces is, predictably, more complicated, but all of the implications applying in normed spaces admit generalisations. This requires the introduction of concepts which differ from (2) in general, but agree with it in the normed case.

For a closed wedge P in a complete metrizable space X, condition (1) is equivalent to order-intervals being bounded, and condition (2) to P generating X. These facts yield some "automatic continuity" results (section 3.5).

The content of those sections which do not form part of the general scheme outlined above will be clear from their titles, except perhaps for section 3.11, which is devoted to characterising linear subspaces E for which E^o is order-convex.

The reader is referred to the explanation of terminology and notation relating to topological linear spaces given in section 0.3. By an "ordered topological linear space" we mean a linear space with a topology and an ordering (both, of course, compatible with the linear structure), but without any connection assumed between the topology and the ordering.

3.1. Elementary theory

We start by listing some elementary facts concerning closed sets and closure. The closure of a wedge is a wedge, but the closure of a cone need not be a cone, as is shown by the lexicographic ordering of R^2. If the positive wedge P is closed, then so are all order-intervals. More generally, so is the order-convex cover of a compact set A, since A + P and A - P are closed (0.3.1). Since a closed set is lineally closed, and the closure of a set contains its lineal closure, 1.3.4 and 1.3.5 give:

(i) a closed wedge is Archimedean;

(ii) if $\overline{P}$ is a cone, then P is almost Archimedean.

The converses are not true: for instance, the cone of finite, non-negative sequences in m is Archimedean, but not closed with respect to the usual topology.

3.1.1. If P is a wedge, and $P \cap U$ is closed for some U in $\mathfrak{N}(X)$, then P is closed.

Proof. Take x not in P. There exists $\lambda > 0$ such that $x \in \mathrm{int}(\lambda U)$. Then $P \cap (\lambda U) = \lambda(P \cap U)$, which is closed. Hence $\mathrm{int}(\lambda U) \sim P$ is an open set containing x.

3.1.2. If a closed cone P exists in X, then the topology of X is Hausdorff.

Proof. If $x \neq 0$, then $x \notin P$ or $-x \notin P$. In either case, there is a neighbourhood of x which does not contain 0.

Order-convexity and closure. Consider R^2 with the usual topology and order. Let

$$A = \{(\xi,\eta) : \xi > 0 \text{ and } \xi\eta = -1\}.$$

Then

$$[A] = \{(\xi,\eta) : \xi > 0 \text{ and } \eta < 0\},$$

so A is closed, while [A] is not. Also, let

$$B = \{(0,\eta) : 0 \leq \eta \leq 1\} \cup \{(1,\eta) : -1 \leq \eta < 0\}.$$

Then B is order-convex, while $\bar{B}$ is not.

Interior points

Any interior point of a subset of a topological linear space is an internal point. Therefore if P is a wedge, any interior point of P is an order-unit with respect to the associated ordering. If A is convex and has interior points, then every internal point of A is an interior point. A sufficient condition for this to occur is that X should be barrelled and A closed. Hence if P is a closed wedge in a barrelled space, then the set of order-units coincides with the set of interior points of P.

3.1.3. Let X be an ordered topological linear space with positive wedge P. Then $e \in \text{int } P$ iff $[-e,e] \in \mathcal{N}(X)$.

Proof. Suppose that $e \in \text{int } P$. Then there exists symmetric U in $\mathcal{N}(X)$ such that $e + U \subseteq P$. Then $U \subseteq (P - e) \cap (e - P) = [-e,e]$.

Conversely, if $[-e,e] = U \in \mathcal{N}(X)$, then $e - U \subseteq P$, so $e \in \text{int } P$.

3.1.4. COROLLARY. If the positive wedge P has an interior point, then every order-bounded linear functional is continuous.

Examples. The usual positive cone on m has $(1,1, \ldots)$ as an interior point. The usual positive cones in s, c_0 and l_1 have no internal points (cf. sect. 1.3).

With respect to the usual norm for l_1, the cone P_s has e_1 as an interior point. However, e_1 is not an interior point of P_s with respect to the supremum norm p, for if

$$x_n = \frac{1}{n}(e_1 + \ldots + e_{n+1}),$$

then $p(x_n) = \frac{1}{n}$ and $e_1 - x_n \notin P_s$.

Linear functionals

For locally convex spaces, we can give topological versions of some of the results of Chapter 1. In the following, polars of subsets of X will be taken in the space X*, so P^0 will mean the set of continuous monotonic linear functionals on X.

3.1.5. Let P be a wedge in a locally convex space X. Then P^0 is a cone iff $P - P$ is dense in X.

Proof. Obvious.

Replacing U by a neighbourhood of 0 in the proof of 1.6.5, we obtain:

<u>3.1.6</u>. Let P be a wedge in a locally convex space X. If A is a convex subset of X, and $x \notin \overline{A - P}$, then there exists f in P^o such that $f(x) > \sup f(A)$.

<u>3.1.7</u>. Let P be a wedge in a locally convex space X. Then:

(i) $P^{oo} = \overline{P}$,

(ii) $\overline{P}$ is a cone iff P^o separates points of X.

<u>Proof</u>. (i) By 3.1.6, if $x \notin \overline{P}$, then there exists f in P^o such that $f(x) < 0$.

(ii) If $\overline{P}$ is a cone and $x \neq 0$, then $x \notin \overline{P}$ or $-x \notin \overline{P}$, so P^o separates points of X, as in (i). If P^o separates points of X, and $x \in \overline{P} \cap (-\overline{P})$, then $f(x) = 0$ for all f in P^o, so $x = 0$.

<u>3.1.8</u>. Let X be an ordered locally convex space with positive wedge P, and suppose that $e \in \operatorname{int} P$. Let E be a linear subspace containing e, and f a monotonic linear functional defined on E. Then f has a continuous, monotonic extension to X.

<u>Proof</u>. The existence of a monotonic extension follows from 1.6.2, and its continuity from 3.1.4.

Replacing U by a neighbourhood of 0 in 1.6.4, we obtain:

<u>3.1.9</u>. Let X be an ordered locally convex space with positive wedge P, and let E be a linear subspace of X. If f is a continuous, monotonic linear functional defined on E, then the following statements are equivalent:

(i) f has a continuous, monotonic linear extension to X;

(ii) there exists U in $\mathfrak{N}(X)$ such that $f \leq 1$ on $E \cap (U - P)$.

To "impose continuity" on 1.6.1, we need a topological version of "cofinal". The next result shows how to do this. As with 1.6.1, we give two different proofs.

3.1.10. Let X be an ordered locally convex space with positive wedge P. Suppose that E is a linear subspace such that for each U in Ⓝ(X), $(E \cap U) - P$ is in Ⓝ(X). Then a continuous, monotonic linear functional defined on E has a continuous, monotonic extension to X.

Proof 1. There exists U in Ⓝ(X) such that $f \leq 1$ on $E \cap U$. Let $V = E \cap U - P$. Then $V \in$ Ⓝ(X), and

$$E \cap (V - P) = E \cap V = E \cap U - E \cap P,$$

so $f \leq 1$ on $E \cap (V - P)$, and 3.1.9 applies.

Proof 2. The first proof of 1.6.1 applies and gives a continuous linear functional, since if

$$p(x) = \inf \{f(y) : y \in E \text{ and } y \geq x\},$$

and $f \leq 1$ on $E \cap U$, then $p(x) \leq 1$ for x in $E \cap U - P$.

We notice that 3.1.8 is a special case of 3.1.10.

Letting U be a neighbourhood of 0 in the proof of 1.7.2, we obtain:

3.1.11. Let X be an ordered locally convex space with positive wedge P, and let f be an element of X. Then the following statements are equivalent:

(i) $f \in P^{o} - P^{o}$.

(ii) There exists U in Ⓝ(X) such that f is bounded above on $P \cap (U - P)$.

Extremal monotonic functionals on locally directed spaces

In order to derive a topological version of 1.8.2, we consider spaces in which the upward-directed neighbourhoods of 0 form a local base. Let us say that a space is locally directed if this condition holds. It is obviously necessary and sufficient if the downward-directed neighbourhoods of 0 form a local base. Examples of such spaces are s, m, c_o and $C[0,1]$, with the usual topologies and orderings.

We shall take a closer look at linear lattices with this property in section 4.3.

If a space is locally convex and locally directed, then 1.1.9(ii) shows that the convex, upward-directed neighbourhoods of 0 form a local base.

<u>3.1.12 THEOREM</u>. Let X be an ordered linear space with a locally convex, locally directed topology. Denote by P the positive wedge in X, by P^o the set of all monotonic linear functionals on X, and by E the set of extremal points of P^o that are continuous. Let A be a subset of X. Then:

(i) If A is directed upwards and $x \notin \overline{A - P}$, then there exists f in E such that $f(x) > \sup f(A)$.

(ii) If A is directed downwards and $x \notin \overline{A + P}$, then there exists f in E such that $f(x) < \inf f(A)$.

<u>Proof</u>. (ii) follows from (i) by considering -x and -A, so we prove (i). Under the conditions of (i), A - P is convex, by 1.1.9(iii), and there is a convex, upward-directed neighbourhood U of 0 such that x - U is disjoint from A - P. Take a_o in A, and let $K = U + (A - a_o) - P$. Then K is convex and upward-directed, and $K - P = K$. Also, $x - a_o \notin K$. Let p be the Minkowski functional of K. Then $p(x - a_o) \geq 1$, so, by 1.8.2, there exists f in E such that $f(x - a_o) \geq 1$. Since f is non-zero, there exists u in U such that $f(u) > 0$. Write $f(u) = \delta$. For a in A, we have

$$f(u + a - a_o) = \delta + f(a) - f(a_o) \leq 1,$$

or $f(a) - f(a_o) \quad 1 - \delta$. Hence $f(x) - f(a) \geq \delta$ $(a \in A)$.

<u>3.1.13 COROLLARY</u>. Under the conditions of 3.1.12, $E^o = \overline{P}$, and if $\overline{P}$ is a cone, then E separates points of X.

<u>Proof</u>. Similar to 3.1.7.

<u>Upper bounds and suprema</u>

We finish this section with some results which apply to ordered topological linear space, either of the following statements is equivalent to the positive wedge being closed:

(1) The sets $\{x : x \geq a\}$ and $\{x : x \leq a\}$ are closed for each a.

(2) The set $\{(x,y) : x \leq y\}$ is closed in $X \times X$.

An ordering of a topological space for which (1) is true will be called a closed ordering. ((2) is in general a stronger condition than (1), and is taken as the definition of a "closed order" in Nachbin (1).) With respect to a closed ordering, a set and its closure have the same upper bounds.

Let $\{x_n : n \in D\}$ be a directed net in an ordered set X. We say that the net is increasing if $x_m \leq x_n$ (in the ordering of X) whenever $m \leq n$ (in the ordering of D). The most natural example of an increasing net is an upward-directed set A, where the ordering of X is used to direct A. If X has a topology, then such a set A converges to x_0 if, given a neighbourhood U of x_0, there exists a_0 in A such that if $a \in A$ and $a \geq a_0$, then $a \in U$.

3.1.14. Let X be a topological space with a closed ordering. If $\{x_n\}$ is an increasing net in X which converges to x_0, then $x_0 = \sup \{x_n\}$.

Proof. For each n, $x_m \geq x_n$ for $m \geq n$, so $x_0 \geq x_n$. If $y \geq x_n$ for all n, then $y \geq x_0$. Hence $x_0 = \sup \{x_n\}$.

However, it is easy to give an example of an increasing sequence which has a supremum but does not converge to it. In m, let $f_n = e_1 + \dots + e_n$. Then $\{f_n\}$ is increasing and $\sup \{f_n\} = e$, but $\|e - f_n\| = 1$ for all n.

3.1.15. Let X be a topological space with a closed ordering, and let A be an upward-directed subset such that, for some a_0 in A, the closure of $\{x \in A : x \geq a_0\}$ is compact. Then $\bar{A}$ contains a supremum of A.

Proof. For each a in A, let B(a) be the closure of $\{x \in A : x \geq a\}$. Since A is directed upwards, the sets B(a) $(a \in A)$ have the finite intersection property. Since one of them is compact, it follows that there is an element x_0 in $\bigcap\{B(a) : a \in A\}$. Then $x_0 \in \bar{A}$ and $x_0 \geq A$. If $y \geq A$, then $y \geq \bar{A}$, so $y \geq x_0$.

3.1.16. COROLLARY. Let X be a topological space with a closed antisymmetric ordering, and let A be a compact subset of X. Given a in A, there exists b in A such that $b \geq a$ and b is a maximal element of A.

Proof. By 3.1.15, each totally ordered subset of A has a supremum in A. The result follows, by Zorn's lemma.

3.1.17 (Nachbin). Let X be a topological space with a closed ordering. Suppose that the Riesz interpolation property holds, and that each order-interval in X is compact. Then X is order-complete.

Proof. Let A be a majorised set, and let B be the set of upper bounds of A. If $x \in B$, then $\{y \in B : y \leq x\}$ is equal to $\bigcap\{[a,x] : a \in A\}$, so is compact. We show that B is directed downwards; it then follows, by 3.1.15, that B has a least element. Take x,y in B. For each a in A, let $C(a) = [a,x] \cap [a,y]$. Then the sets C(a) are compact and, because of the Riesz condition, have the finite intersection property. Hence there is an element z in $\bigcap\{C(a) : a \in A\}$. Then $z \in B$ and $z \leq x$, $z \leq y$.

3.2. Locally order-convex spaces and allied sets

An ordered topological space is said to be locally order-convex if the order-convex neighbourhoods of each point form a base of its neighbourhoods. For a linear space, it is clearly sufficient if this statement holds for the zero element. Local order-convexity is a very natural condition to impose, and we shall see that it has considerable consequences.

Firstly, we notice that if X is locally order-convex and $0 \leq x \leq 0$, then x is in every neighbourhood of 0. If the topology is Hausdorff, the ordering is therefore antisymmetric.

It is clear that the spaces s, m, c_0, C(S), with the usual topologies and orderings, are locally order-convex.

3.2.1. If an ordered topological linear space is locally convex and locally order-convex, then the convex, symmetric, order-convex neighbourhoods of 0 form a local base.

Proof. Take order-convex U in $\mathfrak{N}(X)$. There is a convex, symmetric neighbourhood V contained in U. Then [V] is convex and symmetric, by 1.2.3, and $[V] \subseteq U$.

1.7.3 now yields immediately:

3.2.2. If an ordered topological linear space X is locally convex and locally order-convex, then each continuous linear functional on X is the difference between two continuous, monotonic linear functionals.

We now have a closer look at the meaning of local order-convexity. It is natural to do so in the context of commutative topological groups. As usual, there is an equivalent statement about the positive set P:

3.2.3. Let X be an ordered commutative topological group with positive set P. The following statements are equivalent:

(i) X is locally order-convex;

(ii) given U in $\mathfrak{N}(X)$, there exists V in $\mathfrak{N}(X)$ such that if $x,y \in P$ and $x + y \in V$, then $x,y \in U$.

Proof. (i) => (ii). Given U in $\mathfrak{N}(X)$, there is an order-convex neighbourhood V contained in U. If $x,y \in P$ and $x + y \in V$, then $0 \leq x \leq x + y \in V$, so $x \in V$. Similarly, $y \in V$.

(ii) => (i). Given U in $\mathfrak{N}(X)$, take U_1 in $\mathfrak{N}(X)$ such that $U_1 + U_1 \subseteq U$. There exists V_1 in $\mathfrak{N}(X)$ such that $V_1 \subseteq U_1$ and if $p,p' \in P$ and $p + p' \in V_1$, then $p,p' \in U_1$. Take V in $\mathfrak{N}(X)$ such that $V - V \subseteq V_1$. Suppose that $x \in [V]$, so that there exist v,v' in V and p,p' in P such that $x = v + p = v' - p'$. Then $p + p' = v' - v \in V_1$, so $p \in U_1$.

Hence $x \in V + U_1 \subseteq U$. Thus $[V] \subseteq U$.

The relation between U and V in statement (ii) can be expressed in order notation as follows: if $0 \leq x \leq y \in V$, then $x \in U$.

The property of P expressed in (ii) is a special case of a relation between pairs of subsets of a topological group, which we now define. Let A,B be subsets of a commutative topological group X. We say that A is <u>allied</u> to B, and write A al B, if, given U in $\mathfrak{N}(X)$, there exists V in $\mathfrak{N}(X)$ such that if $a \in A$, $b \in B$ and $a + b \in V$, then $a,b \in U$. We write A nal B in the contrary case. If A al A, we say that A is <u>self-allied</u>. 3.2.3 says that X is locally order-convex iff P is self-allied. (The word "normal" is widely used in the literature instead of "self-allied", but we shall refrain from doing so in accordance with a policy of avoiding overworked words).

A detailed treatment of allied sets is given in Jameson (2). Here we restrict ourselves to results that are relevant to our purposes, but we shall state them in terms of pairs of sets where this gives no extra trouble.

Obviously, if A al B and $A' \subseteq A$, $B' \subseteq B$, then A' al B'.

<u>3.2.4</u>. Suppose that A al B. Then:

(i) Given U in $\mathfrak{N}(X)$, there exists V in $\mathfrak{N}(X)$ such that $\overline{V + A} \cap \overline{V - B} \subseteq U$.

(ii) $\overline{A}$ al $\overline{B}$.

<u>Proof</u>. (i) Take U in $\mathfrak{N}(X)$. As in the proof of 3.2.3, we show that there exists W in $\mathfrak{N}(X)$ such that $(W + A) \cap (W - B) \subseteq U$. Take V in $\mathfrak{N}(X)$ such that $V + V \subseteq W$. Then $\overline{V + A} \subseteq V + V + A \subseteq W + A$, $\overline{V - B} \subseteq W - B$.

(ii) follows, since $V + \overline{A} \subseteq \overline{V + A}$.

Consequently, if P is a self-allied cone in a Hausdorff linear space, then $\overline{P}$ is a cone, and the associated ordering is almost Archimedean.

Alliedness can be characterised in terms of directed nets as follows:

3.2.5. The following condition is equivalent to A al B: if $\{a_n : n \in D\}$ and $\{b_n : n \in D\}$ are directed nets in A,B respectively such that $\{a_n + b_n\}$ converges to 0, then $\{a_n\}$ and $\{b_n\}$ converge to 0. If the topology is metrizable, it is sufficient if this condition holds for sequences.

Proof. Necessity of the condition is obvious. To prove sufficiency, suppose that A nal B. Then there exists U in $\mathcal{N}(X)$ such that for each V in $\mathcal{N}(X)$, there exist a_V in A, b_V in B such that $a_V + b_V \in V$ and one of a_V, b_V is not in U. Let $\mathcal{B}$ be a local base, countable and contracting in the metrizable case, and order $\mathcal{B}$ by inclusion. Then $\{a_V + b_V : V \in \mathcal{B}\}$ is a net convergent to 0, but $\{a_V : V \in \mathcal{B}\}$ and $\{b_V : V \in \mathcal{B}\}$ do not converge to 0.

The order-convex cover

Returning to consequences of local order-convexity, we have:

3.2.6. If X is an ordered linear space with a locally order-convex topology, and A is a bounded subset of X, then [A] is bounded.

Proof. Take order-convex U in $\mathcal{N}(X)$. There exists $\lambda > 0$ such that $\lambda A \subseteq U$. Then $\lambda [A] \subseteq U$.

In particular, order-intervals are bounded. If, in addition, P has an interior point e, then [-e,e] is a neighbourhood of 0, so the topology is semi-normable. (We come back to this in section 3.7).

The converse of 3.2.6 applies in metrizable spaces:

3.2.7. Let X be an ordered metrizable linear space, and suppose that for each bounded subset A of X, $[A]$ is bounded. Then X is locally order-convex.

Proof. Suppose that X is not locally order-convex. Let $\{U_n\}$ be a countable, contracting local base. There exist U in $\mathfrak{N}(X)$ and, for each n, x_n' in $P \sim U$, y_n' in P with $x_n' + y_n' \in n^{-1}U_n$. Let $x_n = nx_n'$. Then $x_n + y_n \to 0$, so if $E = \{0\} \cup \{x_n + y_n\}$, then E is bounded (in fact, compact). However, $[E]$ is unbounded, since $x_n \in [E]$ for each n, and $x_n \notin nU$.

Normed spaces

The argument used in 3.2.1 shows that if a normed linear space is locally order-convex, then there is an equivalent norm whose closed (or open) unit ball is order-convex. For such a norm, $u \leq x \leq v$ implies that $\|x\| \leq \|u\| \vee \|v\|$. Naturally, this condition is more than sufficient for a norm topology to be locally order-convex. In fact, we have:

3.2.8. Let X be an ordered normed linear space with positive wedge P and closed unit ball U. Then the following statements are equivalent:

(i) P is self-allied;

(ii) $[U]$ is bounded;

(iii) there exists $\delta > 0$ such that for x,y in P,

$$\|x + y\| \geq \delta \|x\|.$$

Proof. (i) implies (ii), by 3.2.6, and it is clear that (iii) implies (i). To show that (ii) implies (iii), suppose that $\sup \{ \|x\| : x \in [U]\} = \alpha$. Take non-zero elements x,y of P. Since P is obviously a cone, $\|x + y\| = \mu > 0$. Then

$$0 \leq \mu^{-1}x \leq \mu^{-1}(x + y) \in U,$$

so $\|\mu^{-1}x\| \leq \alpha$. Hence $\|x\| \leq \alpha\|x + y\|$.

In order notation, condition (iii) says that $0 \leq x \leq y$ implies $\|y\| \geq \delta\|x\|$. In particular, it is sufficient if $0 \leq x \leq y$ implies $\|x\| \leq \|y\|$. A case where this is satisfied, while U is not order-convex, is l_1 with the usual topology and order: here we have $(-1,0) \leq (-1,1) \leq (0,1)$, and $\|(-1,1)\| = 2$.

Monotonic nets

The next result is due to Bonsall, its simple proof to Weston.

3.2.9. Let X be an ordered linear space with a locally order-convex topology, and let $\{x_n\}$ be a decreasing net in X. If $x_0 \leq x_n$ for all n and x_0 is in the closure of co$\{x_n\}$, then $\{x_n\}$ converges to x_0.

Proof. Let $y_n = x_n - x_0$. Take order-convex U in Ⓝ(X). There exist a finite set of indices n_i and corresponding numbers λ_i in $(0,1]$ such that $\Sigma\lambda_i = 1$ and $\Sigma\lambda_i y_{n_i} \in U$. If $n \geq n_i$ for each i, then

$$0 \leq y_n = \Sigma\lambda_i y_n \leq \Sigma\lambda_i y_{n_i},$$

so $y_n \in U$. Hence $y_n \to 0$.

3.2.10 COROLLARY. Let τ be a locally convex, locally order-convex topology, and let σ be the associated weak topology. Then every monotonic net that is convergent with respect to σ is convergent with respect to τ.

Proof. Suppose that $\{x_n\}$ is a decreasing net with σ-limit x_0. Convex sets have the same closure with respect to σ and τ. The net is monotonic with respect to the ordering associated with $\overline{P}$, and $x_n - x_0 \in \overline{P}$ for all n. Hence x_0 is the τ-limit of $\{x_n\}$.

Continuity of linear mappings

First we give an obvious generalisation of 3.1.4:

3.2.11. Let X,Y be ordered topological linear spaces with positive wedges P,Q respectively. Suppose that P has an interior point and that Q is self-allied. Then every order-bounded linear mapping from X to Y is continuous.

Proof. Let e be an interior point of P. By 3.1.3, $[-e,e] \in \mathfrak{N}(X)$. If f is order-bounded, then $f[-e,e]$ is contained in an order-interval, so is bounded, by 3.2.6. Hence f is continuous.

If X is a Mackey space and Y is a locally convex space, then a linear mapping f from X to Y is continuous iff $h \circ f$ is continuous for each in in Y^* (see, e.g. Schaefer (4), p. 158). Using this, we have:

3.2.12. Suppose that X,Y are ordered locally convex spaces, X being a Mackey space on which every monotonic linear functional is continuous, and Y being locally order-convex. Then every monotonic linear mapping from X to Y is continuous.

Proof. Take h in Y^*. By 3.2.2, there exist monotonic h_1, h_2 in Y^* such that $h = h_1 - h_2$. Then $h_1 \circ f$ is a monotonic linear functional on X, so is continuous. Hence $h \circ f$ is continuous.

Different topologies

If a cone is self-allied with respect to one topology, is it self-allied with respect to another one? A partial answer to this question can be deduced from 3.2.6 and 3.2.7:

3.2.13 THEOREM. Suppose that X is an ordered linear space, and that τ_1, τ_2 are topologies for X giving the same bounded sets. Suppose, also, that τ_1 is metrizable. If τ_2 is locally order-convex, then so is τ_1.

Proof. If τ_1 is not locally order-convex, then, by 3.2.7, there exists a bounded subset A of X such that $[A]$ is unbounded. By 3.2.6, it follows that τ_2 is not locally order-convex.

One case in which 3.2.13 applies is when τ_1 is a locally convex, metrizable topology and τ_2 is another topology of the same dual pair. Also, the uniform boundedness theorem gives:

3.2.14 COROLLARY. Let X^* be an ordered Banach dual space. If the topology $\sigma(X)$ for X^* is locally order-convex, then so is the norm topology.

We show later (3.4.3) that if a locally convex topology is locally order-convex, then so is the associated weak topology. We will also see (3.5.4) that the converse of 3.2.14 is true whenever X is a Fréchet space.

The topology $F(\tau)$

Let τ be a topology for a linear space X, and let Ⓝ be the family of order-convex τ-neighbourhoods of 0. If U is order-convex and $V + V \subseteq U$, then $[V] + [V] \subseteq U$, so Ⓝ is the local base for a linear topology on X, which is clearly the largest locally order-convex topology contained in τ. Denote this topology by $F(\tau)$. If Ⓑ is a local base for τ, then $\{[U] : U \in Ⓑ\}$ is clearly a local base for $F(\tau)$. Hence if τ is locally convex or pseudo-metrizable, then so is $F(\tau)$. If a set A is τ-bounded, then [A] is $F(\tau)$-bounded, and from this we see that if τ is semi-normable, then so is $F(\tau)$. The question of whehter $F(\tau)$ is Hausdorff can be answered quite neatly as follows:

3.2.15. P has the same closure with respect to τ and $F(\tau)$.

Proof. Suppose that x is in the $F(\tau)$-closure of P. Then, for each τ-neighbourhood U of 0, $x + [U]$ meets P, so $x + U - P$ meets P. Hence $x + U$ meets P, and x is in the τ-closure of P.

3.2.16 COROLLARY. $F(\tau)$ is Hausdorff iff the τ-closure of P is a cone.

Proof. This follows from 3.1.2 and 3.2.4.

Thus if there is a topology (necessarily Hausdorff, by 3.1.2) with respect to which the closure of P is a cone, then there is a

Hausdorff topology with respect to which P is self-allied.

In particular, we can define the largest locally convex, locally order-convex topology. A local base for this is the family of all convex, order-convex, absorbent sets.

3.2.17. Let τ be a locally convex topology for an ordered linear space X, and let P^o denote the set of monotonic, τ-continuous linear functionals on X. Then $P^o - P^o$ is the set of all $F(\tau)$-continuous linear functionals.

Proof. It is clear that each element of P^o, and hence of $P^o - P^o$, is $F(\tau)$-continuous. Conversely, if f is $F(\tau)$-continuous, then, by 3.2.2, f is the difference between two monotonic, $F(\tau)$-continuous linear functionals. Hence $f \in P^o - P^o$.

An example

The "partial-sum cone" P_s is not self-allied in m, c_o or F with respect to the supremum norm. To show this, let

$$x_n = e_1 + \dots + e_n - ne_{n+1},$$
$$y_n = ne_{n+1}.$$

Then $x_n, y_n \in P_s$, $\|x_n\| = \|y_n\| = n$, and $\|x_n + y_n\| = 1$.

We notice that, in the space m with the P_s- ordering, the order-interval $[0,e]$ contains each x_n, so is unbounded (illustrating 3.2.7). Also, $\{e_n\}$ is a decreasing sequence which converges to 0 in the weak topology but not in the norm topology (cf. 3.2.10).

One can easily show that in any of these spaces or s, P_s is self-allied with respect to the topology of pointwise convergence.

3.3. Open decompositions

Let P be a subset of a commutative topological group X. We say that P gives an open decomposition of X if for each neighbourhood U of of 0, the set $P \cap U - P \cap U$ is also a neighbourhood of 0. Similarly, we say that a pair of subsets P,Q gives an open decomposition if, for each U in $\mathfrak{N}(X)$, $P \cap U - Q \cap U \in \mathfrak{N}(X)$.

In this section, we only mention the (rather trivial) results connected with open decompositions in general topological linear spaces. The deeper results concerning this concept apply to complete, metrizable spaces, and are deferred to section 3.5.

If P gives an open decomposition of X, then so does any set containing P. If X is a topological linear space and P is a wedge giving an open decomposition of X, then it is clear that P generates X. If P is a wedge in a normed linear space X, then P gives an open decomposition of X iff there exists $\alpha > 0$ such that, given x in X, there exist a,b in P such that $x = a - b$ and $\|a\|, \|b\| \leq \alpha\|x\|$.

__3.3.1.__ If X is a topological linear space, and P is a wedge in X with non-empty interior, then P gives an open decomposition of X.

__Proof__. Take open U in $\mathfrak{N}(X)$. The set $G = U \cap (\text{int } P)$ is open and non-empty, so $G - G \in \mathfrak{N}(X)$.

__3.3.2.__ Let X be an ordered commutative topological group with positive set P. If X is locally directed, then P gives an open decomposition of X.

__Proof__. Take U in $\mathfrak{N}(X)$. There exists upward-directed V in $\mathfrak{N}(X)$ such that $V - V \subseteq U$. Take x in V. Then there exists v in V such that $v \geq x$ and $v \geq 0$. Then $v - x \in U$, so $x \in P \cap U - P \cap U$.

The notion of open decomposition has some elementary, but useful, applications to continuity.

__3.3.3.__ Let X,Y be topological linear spaces, and suppose that P is a set that contains 0 and gives an open decomposition of X. If f is a linear mapping from X to Y, and the restriction of f to P is continuous at 0, then f is continuous.

__Proof__. Given V in $\mathfrak{N}(Y)$, take V' in $\mathfrak{N}(Y)$ such that $V' - V' \subseteq V$.

There exists U in $\mathfrak{N}(X)$ such that $f(P \cap U) \subseteq V'$. Then $f(P \cap U - P \cap U) \subseteq V$.

3.3.4. Let X,Y be ordered topological linear spaces, with positive wedges P,Q respectively. Suppose that P gives an open decomposition of X, and that Q is self-allied. If f is a continuous, monotonic linear mapping from X to Y, and $0 \leq g \leq f$, then g is continuous. In other words, the space of continuous linear mappings from X to Y is order-convex in L(X,Y).

Proof. Take V in $\mathfrak{N}(Y)$. There exists order-convex V' in $\mathfrak{N}(Y)$ such that $V' - V' \subseteq V$. There exists U in $\mathfrak{N}(X)$ such that $f(U) \subseteq V'$. If $x \in P \cap U$, then $0 \leq g(x) \leq f(x) \in V'$, so $g(x) \in V'$. Hence $g(P \cap U - P \cap U) \subseteq V$.

In particular, if P gives an open decomposition of X, then X^* is order-convex in X'. If P^o is evaluated in X', every extremal point of $P^o \cap X^*$ is therefore an extremal point of P^o (cf. 3.1.12).

Examples

(i) The "usual" positive cones in s, m, c_o, l_1 give open decompositions with respect to the usual topologies.

(ii) If P is a wedge generating a linear space X, then P gives an open decomposition of X with respect to the largest locally convex topology.

(iii) Consider an infinite-dimensional normed linear space with positive cone P and a continuous linear functional f such that $f(x) \geq \|x\|$ $(x \in P)$ (cf. section 3.8). Let $V = \{x \in X: |f(x)| \leq 1\}$. Then V is a $\sigma(X^*)$-neighbourhood of 0, but $P \cap V - P \cap V$ is bounded, so is not a $\sigma(X^*)$-neighbourhood of 0.

This situation can arise when P gives an open decomposition of X with respect to the norm topology. For instance, let X be l_1 with the usual ordering, and let $f(x) = \sum_{n=1}^{\infty} \xi_n$. (Notice that, in this case, P also gives an open decomposition of X with respect to the topology of pointwise convergence).

This negative result has a variant applying to lattices (4.2.6).

(iv) An example in which X^* is not order-convex in X' is as follows. Consider the space F of finite real sequences, ordered by the cone P_d of decreasing non-negative sequences. The space of all linear functionals on F is isomorphic to the space s of all real sequences, the sequence $\{\eta_r\}$ corresponding to the functional f given by:

$$f(x) = \sum_{r=1}^{n} \xi_r \eta_r, \text{ where } x = (\xi_1, \dots, \xi_n).$$

Writing $Y_r = \eta_1 + \dots + \eta_r$ $(1 \leq r \leq n)$, we have

$$f(x) = \xi_1 Y_1 + \sum_{r=2}^{n} \xi_r (Y_r - Y_{r-1})$$

$$= \sum_{r=1}^{n-1} (\xi_r - \xi_{r+1}) Y_r + \xi_n Y_n.$$

It follows that $P_d^o = P_s$. If F is given the l_1-topology, then $F^* = m$. We have seen (section 2.2, example (iv)) that m is not order-convex in s with respect to the P_s-ordering (cf. 3.3.3).

The next result gives a partial characterisation of open decomposition in terms of nets.

<u>3.3.5</u> (Nachbin). Let P be a subset of a commutative topological group X, and consider the statements:

(i) If $\{x_n\}$ is a net convergent to 0, then there exist, for each n, a_n and b_n in P such that $x_n = a_n - b_n$ and the nets $\{a_n\}, \{b_n\}$ converge to 0.

(i)(s) Statement (i), with "net" replaced by "sequence".

(ii) P gives an open decomposition of X.

Then (i) implies (ii), and if the topology is metrizable, (i)(s) is equivalent to (ii).

<u>Proof</u>. Suppose that there exists U in Ⓝ(X) such that $K = P \cap U - P \cap U \notin$ Ⓝ(X). Let Ⓑ be a local base, countable and contracting in the metrizable case. For each V in Ⓑ, take x_V in $V \sim K$. Then $\{x_V : V \in$ Ⓑ$\}$ is a net convergent to 0 (a sequence in the metrizable

case). If $x_V = a_V - b_V$, where $a_V, b_V \in P$, then one of a_V, b_V is not in U. Hence neither of the nets $\{a_V\}$, $\{b_V\}$ converges to 0.

Conversely, it is easily seen that if the topology is metrizable, then (ii) implies (i)(s).

Bounded decompositions

Suppose that X is a commutative group, Ⓐ is a family of subsets of X, and P is a subset of X. We say that P gives an Ⓐ-decomposition of X if, given A in Ⓐ, there exists B in Ⓐ such that $A \subseteq P \cap B - P \cap B$.

If X is a topological linear space, and Ⓐ is the family of bounded subsets of X, then an Ⓐ-decomposition of X will be called a bounded decomposition. The following condition is clearly equivalent: given a bounded set A, there exist bounded subsets B,C of P such that $A \subseteq B - C$.

A wedge in a normed linear space gives a bounded decomposition iff it gives an open decomposition. This equivalence does not hold generally, because bounded decompositions, unlike open decompositions, are the same with respect to all the topologies of a dual pair. However, the implication one way can be extended at least to bornological spaces:

3.3.6. If X is a bornological space, and P is a wedge that gives a bounded decomposition of X, then P gives an open decomposition of X.

Proof. Take convex U in Ⓝ(X), and let $V = P \cap U - P \cap U$. Then V is convex and circled. Take a bounded set A. There is a bounded subset B of P such that $A \subseteq B - B$. For some $\lambda > 0$, $\lambda B \subseteq U$. Then $\lambda A \subseteq V$. Since X is bornological, it follows that $V \in$ Ⓝ(X).

Bounded decompositions share some of the properties of open decompositions: for instance, a wedge with non-empty interior gives a bounded decomposition, and if P gives a bounded decomposition of X, then the set of bounded linear functionals on X is order-convex in X'.

3.4. The duality of self-alliedness and decomposition properties

Let P be a wedge in a linear space X, and let P^o be evaluated in X'. Then we know (cf. 1.5.1) that

(i) if $P - P = X$, then P^o is a cone;

(ii) if P^o is a cone, then $P - P = X$.

This section is concerned with what may be regarded as topological versions of these facts. Results corresponding to implication (i) are basically elementary, and will be dealt with first. Those corresponding to implication (ii) are non-trivial, and depend on the decomposition theorems of 1.7.

3.4.1. THEOREM. If (X,Y) is a dual pair, and P is a wedge generating X, then P^o is self-allied with respect to $\sigma(X)$.

Proof. Take $x_1, \ldots, x_n$ in X. There exist a_i, a_i', b_i, b_i' in P such $x_i = a_i - b_i$, $-x_i = a_i' - b_i'$ $(i = 1, \ldots, n)$. Suppose that $f, g \in P^o$ and that $f + g \leq 1$ at each of a_i, a_i', b_i, b_i'. Then

$$g(a_i) \leq f(a_i) + g(a_i) \leq 1,$$

so

$$g(x_i) = g(a_i) - g(b_i) \leq 1 \quad (i = 1, \ldots, n).$$

Similarly, $g(-x_i) \leq 1$, so $|g(x_i)| \leq 1$.

This theorem, in conjunction with the results of section 3.2, has a number of interesting corollaries.

3.4.2 COROLLARY. Let (X,Y) be a dual pair, and let P be a wedge in X. Then the following statements are equivalent:

(i) P is self-allied with respect to $\sigma(Y)$;

(ii) $P^o - P^o = Y$.

Proof. (i) implies (ii), by 3.2.2. If (ii) holds, then P^{oo} is self-allied with respect to $\sigma(Y)$, by 3.4.1, so (i) holds, since $P \subseteq P^{oo}$.

<u>3.4.3 COROLLARY</u>. Let τ be a locally convex topology for a linear space X, and let σ be the associated weak topology. If a wedge P in X is self-allied with respect to τ, then P is self-allied with respect to σ.

<u>Proof</u>. By 3.2.2, $P^o - P^o = X^*$.

We recall that the converse of 3.4.3 is true if τ is metrizable (3.2.13). The author suspects that this is not necessarily true if τ is not metrizable, but knows of no counter-example.

<u>3.4.4 COROLLARY</u>. If X is a Banach space, and P is a wedge generating X, then P^o is self-allied with respect to the norm topology of X^*.

<u>Proof</u>. Apply 3.4.1 and 3.2.14.

A completely different proof of 3.4.4 is afforded by the methods of section 3.5.

<u>Topologies of uniform convergence</u>

By using the concept of an <u>Ⓐ-decomposition</u> introduced in section 3.3, we can generalise 3.4.1. Let X,Y be topological linear spaces, and let Z be a linear space of continuous linear mappings from X to Y. Let Ⓐ be a family of bounded, symmetric subsets of X such that the union of any two numbers is contained in a third. Then the topology (for Z) of uniform convergence on Ⓐ is defined as follows: sets of the form $\{f : f(A) \subseteq U\}$, where $A \in$ Ⓐ and $U \in$ Ⓝ(Y), constitute a local base. We denote this topology by τ(Ⓐ). If X,Y are normed spaces, and Ⓐ is the family of multiples of the unit ball in X, then τ(Ⓐ) is the norm topology for Z. Under these circumstances, we have the following theorem:

<u>3.4.5 THEOREM</u> (Schaefer). Suppose that X,Y are ordered by positive wedges P,Q where P gives an Ⓐ-decomposition of X and Q is self-allied in Y. Then $\{f \in Z : f(P) \subseteq Q\}$ is self-allied with respect to τ(Ⓐ).

Proof. Take A in Ⓐ and U in Ⓝ(Y). There exists B in Ⓐ such that $A \subseteq P \cap B - P \cap B$. Take order-convex V in Ⓝ(Y) such that $V - V \subseteq U$. Suppose that f and g are monotonic linear mappings in Z, and that $(f + g)(B) \subseteq V$. If $x \in P \cap B$, then

$$0 \leq f(x) \leq (f + g)(x) \in V,$$

so $f(x) \in V$. Hence $f(A) \subseteq V - V \subseteq U$. This proves the result.

We can apply 3.4.5 to linear functionals by setting $Y = R$. Z then becomes a subspace of X^*, and τ(Ⓐ) becomes the polar topology induced on Z by Ⓐ, having local base $\{A^o : A \in$ Ⓐ$\}$. The strong topology and the Mackey topology are obtained by letting Ⓐ be the family of symmetric subsets of X that are (i) bounded or (ii) compact with respect to $\sigma(Z)$. Writing Y for Z in 3.4.5, we obtain:

3.4.6 COROLLARY. Suppose that (X,Y) is a dual pair, and that Ⓐ is a directed family of $\sigma(Y)$-bounded, symmetric subsets of X. If P is a wedge giving an Ⓐ-decomposition of X, then P^o is self-allied with respect to τ(Ⓐ).

We notice that 3.4.1 is the special case arising when Ⓐ is the family of all finite, symmetric subsets of X.

Ⓝ^{o}-decomposition of X^*

It is evident that 3.2.2 does not reproduce the full strength of the decomposition theorem 1.7.3. The concept that enables us to do so (and to derive a converse result from the above theory) is that of an Ⓝ^{o}-decomposition of X^*, where

$$Ⓝ^o = \{U^o : U \in Ⓝ(X)\}.$$

We remind ourselves of the meaning of this: if $P \subseteq X$, then P^o gives an Ⓝ^{o}-decomposition of X^* iff, given U in Ⓝ(X), there exists V in Ⓝ(X) such that $U^o \subseteq P^o \cap V^o - P^o \cap V^o$. An Ⓝ^{o}-decomposition of X^* is the same as a Ⓑ^{o}-decomposition, where Ⓑ is any local base in X. For normed spaces, it is the same as an open (or bounded) decomposition.

3.4.7 THEOREM. Let X be a locally convex space, and write

$$\mathfrak{N}^o = \{U^c : U \in \mathfrak{N}(X)\}.$$

Let P be a wedge in X. Then P is self-allied iff P^o gives an $\mathfrak{N}^o$-decomposition of X^*.

Proof. (i) Suppose that P^o gives an $\mathfrak{N}^o$-decomposition of X^*. Since the topology of X is locally convex, it coincides with $\tau(\mathfrak{N}^o)$. It now follows from 3.4.6 that P is self-allied.

(ii) Suppose that P is self-allied, and take U in $\mathfrak{N}(X)$. There exists convex, symmetric, order-convex V in $\mathfrak{N}(X)$ such that $V \subseteq U$. Then $U^o \subseteq V^o$, and, by 1.7.3, $V^o \subseteq P^o \cap V^o - P^o \cap V^o$.

For locally convex, metrizable spaces, the situation is very tidy:

3.4.8. Let τ be a locally convex, metrizable topology for a linear space X, and let σ be the associated weak topology. Let P be a wedge in X. Then the following statements are equivalent:

(i) P^o gives an $\mathfrak{N}^o$-decomposition of X^*;

(ii) $P^o - P^o = X^*$;

(iii) P is self-allied with respect to σ;

(iv) P is self-allied with respect to τ.

Proof. (i) implies (ii), a priori. (ii) implies (iii), by 3.4.2. (iii) implies (iv), by 3.2.13. (iv) implies (i), by 3.4.7.

An example

We give an example, due to Namioka, where continuous linear functionals can be expressed as a difference of monotonic ones, but not of continuous, monotonic ones. Consider the space F with the usual ordering. Linear functionals on F can be represented by sequences in S (see section 3.3), and each is therefore the difference between two monotonic ones. Let G be the set of elements $(\xi_1,\ldots,\xi_n)$ in F having

$\Sigma\xi_1 = 0$, and let F have the topology $\sigma(G)$. This is a small topology, but it is still Hausdorff. Since we can identify continuous linear functionals on F with elements of G, there are no non-zero monotonic ones.

The order-convex cover of a weakly compact set

3.4.9. Let X be an ordered linear space with a generating positive wedge P. If A is a $\sigma(X)$-compact subset of X', then $[A]$ is $\sigma(X)$-compact. In particular, order-intervals in X' are $\sigma(X)$-compact.

Proof. We use the fact that a subset of X' that is $\sigma(X)$-bounded and $\sigma(X)$-closed is $\sigma(X)$-compact. Now the topology $\sigma(X)$ for X' is locally order-convex, by 3.4.1, so $[A]$ is $\sigma(X)$-bounded, by 3.2.6. Also, $A + P^{o}$ and $A - P^{o}$ are $\sigma(X)$-closed, so $[A]$ is $\sigma(X)$-closed. The result follows.

3.4.10 COROLLARY. Let X be an ordered topological linear space in which the positive wedge P gives an open decomposition. If A is a $\sigma(X)$-compact subset of X^*, then $[A]$ is $\sigma(X)$-compact. In particular, order-intervals in X^* are $\sigma(X)$-compact.

Proof. By 3.3.4, the order-convex cover of A is the same in X^* and X'.

Barrelled spaces

We recall some basic facts about barrelled spaces. Let X be a barrelled space, and let $\beta(X)$ denote the strong topology for X^*. Denote by Ⓑ the set of $\sigma(X)$-bounded subsets of X^*. If B ∈ Ⓑ, then $B^{o} \in$ Ⓝ(X), so B^{oo} is $\sigma(X)$-compact. Also, if A is a bounded subset of X, then B^{o} absorbs A, so A^{o} absorbs B. Hence (i) Ⓑ is also the set of $\beta(X)$-bounded subsets of X^*, (ii) every $\sigma(X)$-closed set in Ⓑ is $\sigma(X)$-compact. Because of (i), a Ⓑ-decomposition of X^* will simply be called a "bounded decomposition". If U ∈ Ⓝ(X), then U^{o} ∈ Ⓑ, so the Ⓝ^{o}-decompositions of X^* are precisely the bounded decompositions, and 3.4.7 can therefore be restated as follows:

3.4.11. Let P be a wedge in a barrelled space X. Then P is self-allied iff P^o gives a bounded decomposition of X^*.

We show now that we can get close to proving a converse result to 3.4.1 for barrelled spaces. We make use of some elementary facts about polars (defined, as always as in 0.2). If (X,Y) is a dual pair, if A is a subset of either X or Y, and $0 \in A$, then it follows at one from the separation theorem that $A^{oo} = \overline{co}(A)$, where the latter denotes the closure of co(A) in the weak topology of the pair. For any A,B, it is clear that $(A \cup B)^o = A^o \cap B^o$. Hence if $0 \in A \cup B$, then $(A^o \cap B^o)^o = \overline{co}(A \cup B)$. Further, if A,B are weakly closed, convex sets containing 0, then $(A \cap B)^o = \overline{co}(A^o \cup B^o)$.

We separate out the following lemma so that it can be used again later.

3.4.12 LEMMA. Let X be an ordered locally convex space with closed positive wedge P, and let U be a closed, convex, symmetric set in $\mathfrak{N}(X)$. Write $Q = P \cap U$. Then

$$Q^o \cap (-Q^o) \subseteq [U^o],$$

where the polars are taken in X^*.

Proof. Since convex subsets of X have the same closure with respect to $\sigma(X^*)$ and the given topology, we have $Q^o = \overline{co}(U^o \cup P^o)$, the bar denoting closure with respect to $\sigma(X)$. Now U^o is $\sigma(X)$-compact, so $U^o + P^o$ is $\sigma(X)$-closed, and therefore contains $\overline{co}(U^o \cup P^o)$. Therefore $Q^o \subseteq U^o + P^o$. Hence also $-Q^o \subseteq U^o - P^o$, and the result follows.

3.4.13. THEOREM. Suppose that X is a barrelled space, and that P is a closed wedge in X. If P^o is self-allied with respect to $\sigma(X)$ or $\beta(X)$, then, for each U in $\mathfrak{N}(X)$, the closure of $P \cap U - P \cap U$ is in $\mathfrak{N}(X)$.

Proof. Take a closed, convex, symmetric set U in Ⓝ(X), and write $Q = P \cap U$. Then U^{o} is bounded (with respect to $\sigma(X)$ and $\beta(X)$), so the same is true of $[U^{o}]$. Therefore, by 3.4.12. $Q^{o} \cap (-Q^{o})$ is bounded, so the polar of this set, i.e. $\overline{co}(Q \cup (-Q))$, is a neighbourhood of 0. The result follows, since $co(Q \cup (-Q)) \subseteq Q - Q$.

The author does not know whether being self-allied with respect to $\beta(X)$ is always equivalent to being self-allied with respect to $\sigma(X)$ (but see 3.5.4 and the comment after it).

3.5. Complete metrizable spaces

Open decomposition

Our first major result on complete metrizable spaces is that any closed wedge generating such a space automatically gives an open decomposition of it. A rather stronger statement is, in fact, true. We prove it in two stages; the first stage applies to commutative groups.

3.5.1. THEOREM. Let X be a complete, metrizable commutative topological group, and let A, B be closed semigroups in X. Let d be an invariant metric inducing the topology, and write

$$p(x) = d(x,0), \quad U_{\alpha} = \{x : p(x) \leq \alpha\} \qquad (\alpha > 0).$$

Suppose that, for each $\alpha > 0$, the closure of $A \cap U_{\alpha} - B \cap U_{\alpha}$ is a neighbourhood of 0. Then, for all $\beta > \alpha$, the set $A \cap U_{\beta} - B \cap U_{\beta}$ contains the closure of $A \cap U_{\alpha} - B \cap U_{\alpha}$, so is a neighbourhood of 0.

Proof. For each $\alpha > 0$, there exists $Q(\alpha) > 0$ such that the closure of $A \cap U_{\alpha} - B \cap U_{\alpha}$ contains $U_{Q}(\alpha)$. Let $\beta = \alpha + \varepsilon$, and take x in the closure of $A \cap U_{\alpha} - B \cap U_{\alpha}$. Then there exist a_1 in $A \cap U_{\alpha}$ and b_1 in $B \cap U_{\alpha}$ such that $p(x_1) \leq Q(\frac{\varepsilon}{2})$, where $x_1 = x - (a_1 - b_1)$. We repeat this construction. Having obtained x_{n-1} such that $p(x_{n-1}) \leq Q(\frac{\varepsilon}{2^{n-1}})$, take a_n in A and b_n in B such that $p(a_n)$ and $b(b_n)$ are not greater than $\varepsilon/2^{n-1}$ and $p(x_n) \leq Q(\frac{\varepsilon}{2^{n}})$, where $x_n = x_{n-1} - (a_n - b_n)$.

Since X is complete and A is a closed semigroup, we have $\sum_{n=1}^{\infty} a_n = a$, where $a \in A$ and

$$p(a) \leq \Sigma p(a_n) < \alpha + \frac{\varepsilon}{2} + \frac{\varepsilon}{2^2} + \dots = \beta.$$

Similarly, $\sum_{n=1}^{\infty} b_n = b \in B \cap U_\beta$. Now $\sum_{r=1}^{n} (a_r - b_r) = x - x_n$, and $x_n \to 0$, so $x = a - b$.

<u>3.5.2. THEOREM</u>. Let X be a complete, metrizable topological linear space. Suppose that A,B are closed wedges such that, given x in X, there exist bounded sequences $\{a_n\}$ in A, $\{b_n\}$ in B such that $a_n - b_n \to x$. Then (A,B) gives an open decomposition of X.

<u>Proof</u>. With the notation of 3.5.1, let $\alpha > 0$ be given, and take circled V in Ⓝ(X) such that $V + V \subseteq U$. Let L be the closure of $A \cap V - B \cap V$, and let $K = L \cap (-L)$. Given x in X, there exist bounded sequences $\{a_n\}$, $\{a'_n\}$ in A and $\{b_n\}$, $\{b'_n\}$ in B such that $a_n - b_n \to x$ and $a'_n - b'_n \to -x$. There exists a positive integer r such that $a_n, a'_n, b_n, b'_n \in rV$ for all n. Then $x \in rK$. By Baire's theorem, it follows that there exist x_0 in X and $\delta > 0$ such that $x_0 + U_\delta \subseteq K$. Take x such that $p(x) < \delta$. We show that x is in the closure of $A \cap U_\alpha - B \cap U_\alpha$, from which the result follows, by 3.5.1. Given W in Ⓝ(X), take W_1 in Ⓝ(X) such that $W_1 + W_1 \subseteq W$. There exist a,a' in $A \cap V$, b,b' in $B \cap V$ and w,w' in W_1 such that

$$x_0 + x = a - b + w,$$
$$-x_0 = a' - b' + w'.$$

Then

$$x = (a + a') - (b + b') + (w + w'),$$

so $x - W$ meets $A \cap U_\alpha - B \cap U_\alpha$.

In this theorem, and others depending on it, it is enough to assume that A and B are <u>series-closed</u> instead of closed, that is, that each contains the sum of every convergent series of its elements.

We have seen (section 3.3, example (iv)) that P_d generates F but does not give an open decomposition of it with respect to the l_1-norm. This shows that 3.5.2 can fail in the absence of completeness.

Self-allied cones

Remembering that Fréchet spaces are barrelled, we can combine 3.4.13 and 3.5.1 to obtain:

3.5.3. Suppose that X is a Fréchet space, and that P is a closed wedge in X. If P^o is self-allied with respect to $\sigma(X)$ or $\beta(X)$, then P gives an open decomposition of X.

3.5.4. COROLLARY. Let X be a Fréchet space, and let Q be a $\sigma(X)$-closed wedge in X^*. If Q is self-allied with respect to $\beta(X)$, then Q is self-allied with respect to $\sigma(X)$.

Proof. By 3.5.3 and 3.4.1.

We know from 3.2.14 that the converse of 3.5.4 is true when X is a Banach space, so that the same $\sigma(X)$-closed wedges in X^* are then self-allied with respect to $\sigma(X)$ and $\beta(X)$.

Continuity

An important consequence of 3.5.2 is the following automatic-continuity theorem, which should be compared with 3.2.11.

3.5.5. THEOREM. Let X,Y be topological linear spaces, X being complete and metrizable. Suppose that X is ordered by a closed, generating wedge P. Let f be a linear mapping from X to Y that maps order-intervals into bounded sets. Then f is continuous.

Proof. Let d be an invariant metric giving the topology of X, and let $U_n = \{x : d(x,\theta) \leq 2^{-n}\}$. By 3.5.2, $n^{-1}(P \cap U_n - P \cap U_n)$ is in Ⓝ(X). Therefore if f is not continuous, there exist V in Ⓝ(Y) and a_n, b_n in $P \cap U_n$ such that $n^{-1}f(a_n - b_n) \notin V$. Since X is complete and

P is closed, $\sum_{n=1}^{\infty} a_n$ converges to an element a, and $0 \leq a_n \leq a$ for all n. Since $f[0,a]$ is bounded, by hypothesis, it follows that $n^{-1}f(a_n) \to 0$. Similarly, $n^{-1}f(b_n) \to 0$. This is a contradiction, and the result follows.

The theorem applies, in particular, if Y is a locally order-convex ordered linear space and f is order-bounded. Specialising to linear functionals, we have:

3.5.6. COROLLARY. Let X be a complete, metrizable linear space with an ordering given by a closed, generating wedge. Then every order-bounded linear functional on X is continuous.

Again, a simple example shows that we cannot dispense with completeness. Consider F, this time with the usual ordering and the supremum norm. The linear functional defined by $f(\{\xi_n\}) = \Sigma\xi_n$ is monotonic but not continuous.

Linear functionals on Fréchet spaces

For Fréchet spaces, we can now prove a decomposition theorem that does not require local order-convexity (cf. 3.4.7).

3.5.7. Let X be an ordered Fréchet space with a closed positive wedge P. If f is a continuous, order-bounded linear functional on X, then there exist continuous, monotonic linear functionals g,h such that f = g - h.

Proof. Let d be an invariant metric giving the topology, and let $U_n = \{x : d(x,0) \leq 2^{-n}\}$. We show that, for some n, f is bounded above on $P \cap (U_n - P)$, from which the result follows, by 3.1.11. If this is not the case, then there exist, for each n, x_n in U_n and y_n such that $0 \leq y_n \leq x_n$ and $f(y_n) > n$. Let $x = \sum_{n=1}^{\infty} x_n$. Then $0 \leq y_n \leq x$ for each n, so f is unbounded on $[0,x]$.

Applying 3.5.6, we now have:

3.5.8. COROLLARY. Let X be a Fréchet space, ordered by a closed, generating wedge. If f is an order-bounded linear functional on X, then f is continuous, and is the difference between two continuous, monotonic linear functionals.

Local order-convexity

In a metrizable space, if the order-convex cover preserves boundedness, then the topology is locally order-convex (3.2.7). A stronger result applies if the space is complete:

3.5.9. THEOREM. Let X be a complete, metrizable topological linear space with an ordering given by a closed wedge P. If all order-intervals are bounded, then X is locally order-convex.

Proof. Let d be an invariant metric giving the topology, and suppose that X is not locally order-convex. Then there exist U in $\mathfrak{N}(X)$ and x_n, y_n such that, for each n, $d(0,x_n) \leq 2^{-n}$, $0 \leq y_n \leq x_n$ and $y_n \notin nU$. Let $x = \sum_{n=1}^{\infty} x_n$. Then $0 \leq y_n \leq x$ for each n, so the order-interval $[0,x]$ is unbounded.

3.5.10. COROLLARY. Let X be a Fréchet space with an ordering given by a closed wedge P. Then X is locally order-convex iff every continous linear functional is order-bounded.

Proof. If every continous linear functional is order-bounded, then, by 0.3.3, every order-interval is bounded.

Consider the space F with the supremum norm. We saw in section 3.2 that the cone P_s is not self-allied. We show now that P_s-order-intervals are bounded, so that 3.5.9 also fails in the absence of completeness. Take $a = (\alpha_1, \ldots, \alpha_k)$ in P_s, and let

$$A_r = \alpha_1 + \ldots + \alpha_r \ (1 \leq r \leq k), \qquad A = \max A_r.$$

If $x = \{\xi_n\}$ and $0 \leq x \leq a$, then $0 \leq X_r \leq A$ for all r in ω, where

$X_r = \xi_1 + \ldots + \xi_r$. Hence $|\xi_r| = |X_r - X_{r-1}| \leq A$ for all r, i.e. $\|x\| \leq A$.

This example also serves to show that the property that order-intervals are bounded (unlike local order-convexity) is not preserved if P is replaced by $\overline{P}$. For we know, by 3.5.8, that in c_o with the P_s-ordering, there is an unbounded order-interval. But $P_s = \overline{P}$, where $P = P_s \cap F$, and a P-order-interval of the form $P \cap (a - P)$ is contained in F, and is therefore order-bounded, as above.

Conditions for completeness

In the converse direction to 3.5.2, we have:

3.5.11. Suppose that X is a metrizable, commutative topological group, and that P is a semigroup giving an open decomposition of X. If each increasing Cauchy sequence in P has a limit, then X is complete.

Proof. Let d be an invariant metric giving the topology, and let $U_n = \{x : d(x,0) < 2^{-n}\}$, $V_n = P \cap U_n - P \cap U_n$. Any Cauchy sequence in X has a subsequence $\{x_n\}$ such that $x_n - x_{n-1} \in V_n$ for each n(where $x_0 = 0$). Take a_n, b_n in $P \cap U_n$ such that $x_n - x_{n-1} = a_n - b_n$ $(n = 1,2,\ldots)$. Then $\sum_{r=1}^{n} (a_r - b_r) = x_n$. By hypothesis, $\sum_{n=1}^{\infty} a_n$ and $\sum_{n=1}^{\infty} b_n$ converge, and hence $\{x_n\}$ does.

We now prove an interesting variant of this result, in which the existence of suprema, not limits, is assumed. This is a first step towards relating order-completeness to (topological) completeness. In general, there is not a strong connection between these properties. For instance, C[0,1] is complete but not order-complete, while F is order-complete but not complete with respect to any metrizable linear topology.

3.5.12. Suppose that X is a metrizable linear space, and that P is a closed, self-allied cone giving an open decomposition of X. If each increasing Cauchy sequence in P has a supremum, then X is complete.

__Proof__. Let $\{x_n\}$ be an increasing Cauchy sequence in P, and let $x = \sup \{x_n\}$. We show that $x_n \to x$, from which the result follows, by 3.5.11. Let d be an invariant metric giving the topology, and let $U_n = \{x : d(x,0) \leq 2^{-n}\}$. Taking a subsequence if necessary, we may suppose that $x_n - x_{n-1} \in n^{-1}U_n$ for each n (where $x_0 = 0$). Let $y_n = \sum_{r=1}^{n} r(x_r - x_{r-1})$. If $m < n$, then

$$y_n - y_m = \sum_{m+1}^{n} r(x_r - x_{r-1}) \in U_{m+1} + \ldots + U_n \subseteq U_m,$$

so $\{y_n\}$ is an increasing Cauchy sequence, and has a supremum, y. Now

$$\begin{aligned} 0 \leq m(x - x_m) &= m \sup_{n>m} (x_n - x_m) \\ &= m \sup_{n>m} \sum_{m+1}^{n} (x_r - x_{r-1}) \\ &\leq \sup_{n>m} (y_n - y_m) \\ &= y - y_m. \end{aligned}$$

Hence $0 \leq x - x_m \leq m^{-1}y$ for all m. Since P is self-allied, it follows that $x_m \to x$.

Further results relating order-completeness and completeness are 3.7.4, 3.8.9. For a comprehensive survey of such results, we refer to Peressini (2), ch. 4.

3.6. Normed spaces

It follows from 3.4.6, 3.4.7 and 3.5.3 that if P is a wedge in a normed space X, then P is self-allied iff P^o gives a bounded decomposition of X^*, and that if P is a closed wedge in a Banach space X, then P gives a bounded decomposition of X iff P^o is self-allied. These topological results can be sharpened into numerical ones in which we use the norm, roughly speaking, to measure how self-allied a cone is, or how bounded a decomposition is. We describe two different ways of doing this, corresponding to the two decomposition theorems 1.7.3 and 1.7.1 (in that order). The closed unit ball will be denoted consistently by U.

First duality theory

Let P be a wedge in a normed linear space X. We say that

(1) P is α-normal if $[U]\_ \subset \alpha U$, i.e. if $\|u\|$, $\|v\| \leq 1$ and $u \leq x \leq v$ imply $\|x\| \leq \alpha$;

(2) P is α-generating if, given x in X, there exist p,q in P such that $x = p - q$ and $\|p\| + \|q\| \leq \alpha\|x\|$.

In both cases, it is clear that we must have $\alpha \geq 1$. P is 1-normal iff U is order-convex: this occurs for the usual orderings of m and c_o, but the usual positive cone in l_1 is 2-normal, not 1-normal (cf. the remark following 3.2.8). The usual cone in l_1 is 1-generating, whereas those in m and c_o are 2-generating.

Before embarking on the duality theory suggested by these considerations, we give what can be regarded as a numberical version of 3.3.3. B(X,Y) denotes the set of bounded linear mappings from X to Y.

3.6.1. Let X,Y be normed linear spaces, and suppose that P is an α-generating wedge in X. For f in B(X,Y), define

$$p(f) = \sup \{ \|f(x)\|: x \in P \text{ and } \|x\| \leq 1 \}.$$

Then p is a norm on B(X,Y), and, for all f in B(X,Y)

$$p(f) \leq \|f\| \leq \alpha p(f).$$

Proof. It is elementary that p is a seminorm and that $p(f) \leq \|f\|$. Take x with $\|x\| \leq 1$. There exist p,q in P such that $x = p - q$ and $\|p\| + \|q\| \leq \alpha$. Thus

$$\begin{aligned}\|f(x)\| &\leq \|f(p)\| + \|f(q)\| \\ &\leq p(f)(\|p\| + \|q\|) \\ &\leq \alpha\, p(f).\end{aligned}$$

The numerical result corresponding to 3.4.7 is:

3.6.2. THEOREM. Let P be a cone in a normed linear space X. Then the following statements are equivalent:

(i) P is α-normal;

(ii) P^o is α-generating in X^*.

Proof. (i) => (ii). Take f in X^* such that $\|f\| = 1$. Let U denote the closed unit ball in X. Then [U] is convex and symmetric, and $\sup f([U]) \leq \alpha$. Therefore, by 1.7.3, there exist monotonic linear functionals g,h on X such that $f = g - h$ and

$$\sup g([U]) + \sup h([U]) \leq \alpha.$$

Then $g,h \in X^*$, and $\|g\| + \|h\| \leq \alpha$.

(ii) => (i). Suppose that $u,v \in U$ and $u \leq x \leq v$. Take f in X^* with $\|f\| = 1$. There exist g,h in P^o such that $f = g - h$ and $\|g\| + \|h\| \leq \alpha$. Then $g(u) \leq g(x) \leq g(v)$, so $|g(x)| \leq \|g\|$. Similarly, $|h(x)| \leq \|h\|$. Hence

$$|f(x)| \leq |g(x)| + |h(x)| \leq \alpha.$$

It follows that $\|x\| \leq \alpha$.

Essentially the same reasoning as that used in the second implication here yields the following result, which can be regarded as the numerical version of 3.4.5.

3.6.3. Let X,Y be normed linear spaces. Suppose that P is an α-generating wedge in X, and that Q is a β-normal cone in Y. Then $\{f : f(P) \subseteq Q\}$ is $\alpha\beta$-normal in B(X,Y).

Proof. Suppose that $f,g_1,g_2 \in B(X,Y)$, $\|g_i\| \leq 1$ $(i = 1,2)$, and $g_1 \leq f \leq g_2$. Take x in P with $\|x\| \leq 1$. Then $g_1(x) \leq f(x) \leq g_2(x)$, so $\|f(x)\| \leq \beta$. Hence, by 3.6.1, $\|f\| \leq \alpha\beta$.

3.6.4. COROLLARY. If P is an α-generating wedge in X, then P^o is α-normal in X^*.

We notice that 3.6.4, combined with 3.5.2, gives an alternative proof of 3.4.4.

Let A be a subset of a topological linear space X. We say that A is CS-closed if the following holds: if sequences $\{a_n\}$ in A and $\{\lambda_n\}$ in [0,1] are such that $\sum_{n=1}^{\infty} \lambda_n = 1$ and $\sum_{n=1}^{\infty} \lambda_n a_n$ converges to an element x of X, then $x \in A$.

Our numerical versions of 3.5.3 will be proved with the aid of the following theorem.

A.1. If A is a CS-closed subset of a metrizable linear space, then A and its closure $\bar{A}$ have the same interior.

The proof of this theorem, together with some elementary facts about CS-closed sets, is given in the Appendix. Any easy corollary is the following result, which generalises a theorem of Tukey:

A.2. Let X be a metrizable linear space. Suppose that A,B are closed, convex subsets of X, and that A is bounded and complete. Then co(A∪B) and its closure $\overline{co}(A \cup B)$ have the same interior.

We now conclude our first duality theory with:

3.6.5. THEOREM (Ellis, Andô). Suppose that P is a closed wedge in a Banach space X. If P^o is α-normal, then, for all $\beta > \alpha$, P is β-generating.

Proof. Let U be the closed unit ball in X, and write $Q = P \cap U$. By 3.4.12,

$$Q^o \cap (-Q^o) \subseteq [U^o] \subseteq \alpha U^o$$

Taking polars, it follows that

$$\overline{co}(Q \cup (-Q)) \supseteq \alpha^{-1} U.$$

A.2 now shows that, for $\beta > \alpha$,

$$co(Q \cup (-Q)) \supseteq \beta^{-1} U.$$

Hence if $x \in U$, there exist p,q in $P \cap U$ and $\lambda, \mu \geqslant 0$ such that

$\lambda + \mu \leq \beta$ and $x = \lambda p - \mu q$. Therefore P is β-generating.

It is possible to give a proof of our basic "forward" decomposition theorem 1.7.3 on somewhat similar lines (see Riedl (1), p. 100).

In 3.6.5 it is clearly sufficient if P is series-closed (cf. the comment following 3.5.2).

Second duality theory

Our second duality theory is concerned with order-intervals of the form $[-x,x]$, and is better suited than the first theory for applications to normed lattices (see Chapter 4). A similar theory applies to order-intervals of the form $[0,x]$ (see Ng (2)). There is an obvious correspondence between the results of our two duality theories, except that the analogue of 3.6.1 in the second theory only applies to monotonic linear mappings into spaces with a self-allied cone (we simplify matters by considering only linear functionals).

For the purpose of the second duality theory, we shall say that a wedge P in a normed linear space X has:

(a) property (N) with constant α if $-x \leq y \leq x$ implies that $\|y\| \leq \alpha\|x\|$;

(b) property (G) with constant α if, given x in X, there exists y in X such that $-y \leq x \leq y$ and $\|y\| \leq \alpha\|x\|$.

Clearly, P has property (N) for some α iff it is self-allied, and P has property (G) for some α iff it gives a bounded decomposition of X. The usual positive cones in m, c_0 and l_1 have both properties with constant 1.

3.6.6. Let X be a normed linear space with closed unit ball U, and let P be a wedge in X that has property (G) with constant α. Then:

(i) for f in P^0, $\|f\| \leq \alpha \sup f(P \cap U)$.

(ii) P^0 has property (N) with constant α.

Proof. (i) Given $\varepsilon > 0$, there exists x in U such that

$f(x) > \|f\| - \varepsilon$. There exists y in X such that $-y \leq x \leq y$ and $\|y\| \leq \alpha$. Then $f(y) > \|f\| - \varepsilon$.

(ii) Suppose that $f, g \in X^*$ and $-f \leq g \leq f$. Take x in U. There exists y in X such that $-y \leq x \leq y$ and $\|y\| \leq \alpha$. Then $g(x) \leq f(y)$, by 1.5.4, so $g(x) \leq \alpha\|f\|$. Hence $\|g\| \leq \alpha\|f\|$.

<u>3.6.7</u>. Let P be a cone in a normed linear space X. Then the following statements are equivalent:

(i) P has property (N) with constant α;

(ii) P^o has property (G) with constant α.

<u>Proof</u>. (i) => (ii). This follows at once from 1.7.1.

(ii) =>(i). Similar to 3.6.6 (ii).

It remains to prove a result corresponding to 3.5.3. Our proof (which is rather shorter than those previously published) makes use of 1.7.1 and A.1.

<u>3.6.8. THEOREM</u>. Suppose that P is a closed wedge in a Banach space X. If P^o has property (N) with constant α, then, for all $\beta > \alpha$, P has property (G) with constant β.

<u>Proof</u>. Let $V = \bigcup\{[-x,x] : x \in P \cap U\}$. It is elementary that V is CS-closed (and it is sufficient if P is series-closed). If $f \in V^o$, then, by 1.7.1, there is a member g of X^* such that $-g \leq f \leq g$ and $\|g\| \leq 1$. Therefore, by hypothesis, $\|f\| \leq \alpha$. Hence $V^o \subseteq \alpha U^o$, so $V^{oo} = \overline{V} \supseteq \alpha^{-1} U$. By A.1, it follows that $V \supseteq \beta^{-1} U$ for $\beta > \alpha$. This proves the theorem.

<u>3.7. Order-unit seminorms</u>

If e is an order-unit, then the Minkowski functional of $[-e,e]$ is a seminorm, called the <u>order-unit seminorm</u> corresponding to e. Explicitly, this seminorm is $\| \ \|$, where

$$\|x\| = \inf\{\lambda > 0 : -\lambda e \leq x \leq \lambda e\}.$$

Clearly, $\| \ \|$ is a norm iff the ordering is almost Archimedean. We assume throughout that P is a proper subset of X: otherwise $\|x\| = 0$ for all x. By 1.3.2, this assumption implies that $-e \notin P$, and hence that $\|e\| = 1$.

If e,f are order-units, then there exist $\alpha, \beta > 0$ such that $\alpha f \leq e \leq \beta f$, so the order-unit seminorms corresponding to e and f are equivalent. Because of this, it is reasonable to regard the topology given by each of them as a topology induced by the ordering: we shall return to this point in Section 3.10. Here we start by showing that there is a simple answer to the question of when a given topology is that induced by an order-unit seminorm.

3.7.1. Let X be an ordered topological linear space with positive wedge P. Then the following statements are equivalent:

(i) There is a local base consisting of order-intervals.

(ii) P is self-allied and has interior points.

(iii) There is an order-unit seminorm inducing the topology.

Proof. (i) => (ii). Condition (i) clearly implies local order-convexity. If $[a,b] \in \mathfrak{N}(X)$, then $b \in \operatorname{int} P$, since $b - [a,b] \subseteq P$.

(ii) =>(iii). Let e be an interior point of P. Then $[-e,e] \in \mathfrak{N}(X)$, by 3.1.3, and $[-e,e]$ is bounded, by 3.2.6. Hence the scalar multiples of $[-e,e]$ form a local base.

(iii) => (i). Obvious.

The two conditions in 3.7.1(ii) can be considered separately, as follows. Let e be an order-unit, and let τ_e be the topology given by the corresponding seminorm. Let τ be some linear topology for X. If P is self-allied with respect to τ, then $[-e,e]$ is τ-bounded, so $\tau \subseteq \tau_e$. If P has interior points with respect to τ, then $[-e,e]$ is a τ-neighbourhood of 0, so $\tau_e \subseteq \tau$.

Example

Let X be the space of all bounded real-valued functions on a set S, with the usual ordering. Let $e(s) = 1$ $(s \in S)$. Then e is an order-unit, and the corresponding seminorm is $\| \ \|$, where

$$\|x\| = \inf\{\lambda > 0 : -\lambda \leq x(s) \leq \lambda \text{ for all } s \text{ in } S\}$$
$$= \sup\{|x(s)| : s \in S\},$$

so $\| \ \|$ is the usual supremum norm.

If we consider the space m ordered by P_s, then e is again an order-unit, and the corresponding norm $\| \ \|$ is majorised by the supremum norm. It is not equivalent, since $\|e_n\| = n^{-1}$ for each n.

In the next result, we summarise some elementary properties of order-unit seminorms.

3.7.2. Suppose that e is an order-unit, and that $\| \ \|$ is the corresponding seminorm. Let $U = \{x : \|x\| \leq 1\}$. Then:

(i) U is order-convex.

(ii) If the ordering is Archimedean, then $U = [-e,e]$ and P is closed.

(iii) If f is a monotonic linear functional, then $\|f\| = f(e)$.

Proof. (i) Suppose that $u,v \in U$ and $u \leq x \leq v$. For $\lambda > 1$, $-\lambda e \leq u \leq x \leq v \leq \lambda e$. Hence $x \in U$.

(ii) Suppose that the ordering is Archimedean. If $\|x\| \leq 1$, then $-\lambda e \leq x \leq \lambda e$ for all $\lambda > 1$, so $-e \leq x \leq e$, by 1.3.4. If $x \in \overline{P}$, then, for each n, there exists p_n in P such that $\|x - p_n\| < n^{-1}$. Then $x - p_n \geq -n^{-1}e$, so $x \geq -n^{-1}e$. Hence $x \in P$.

(iii) As mentioned above, $\|e\| = 1$, so $\|f\| \geq f(e)$. If $\|x\| \leq 1$, then $-e \leq x \leq e$, so $|f(x)| \leq f(e)$.

An order-unit is an interior point of P with respect to its own seminorm; in fact, $e + x \in P$ whenever $\|x\| \leq 1$. It is immediate, in the terminology of the second duality theory of 3.6, that an order-unit semi-norm makes the positive wedge have property (G) with constant 1. In the terminology of the first duality theory, however, we can only say that P is 3-generating.

We saw in section 1.8 that the existence of order-units implies the existence of extremal monotonic linear functionals. This can easily be related to order-unit seminorms:

3.7.3 Let X be an ordered linear space with a proper positive wedge P. Suppose that e is an order-unit and that $\| \ \|$ is the corresponding seminorm. Then, given x_0 in X, there exists an extremal element f of P^o such that $\|f\| = 1$ and $|f(x_0)| = \|x_0\|$.

Proof. Let $p(x) = \inf \{\lambda > 0 : x \leq \lambda e\}$ $(x \in X)$. It is easily verified that $\|x\| = p(x) \vee p(-x)$ $(x \in X)$. If $\|x_0\| = p(x_0)$, then the result follows by 1.8.3. If $\|x_0\| = p(-x_0)$, then, by 1.8.3, there exists an extremal element f of P^o such that $\|f\| \leq 1$ and $f(-x_0) = p(-x_0)$, or $-f(x_0) = \|x_0\|$.

We finish this section by showing that quite a weak form of order-completeness implies completeness with respect to an order-unit semi-norm:

3.7.4. Let X be an ordered topological linear space in which the positive wedge P is self-allied and has an interior point. If every majorised increasing sequence in X has a supremum, then the topology for X is complete.

Proof. Let e be an interior point of P. By 3.7.1, the topology is given by the order-unit seminorm $\| \ \|$ corresponding to e. By 3.5.11, it is sufficient to show that any increasing Cauchy sequence $\{x_n\}$ has a limit. Given $\varepsilon > 0$, there exists N such that for $m,n \geq N$, $\|x_m - x_n\| < \varepsilon$, so that $x_m \leq x_N + \varepsilon e$ $(m \geq N)$. Hence $\{x_n\}$ is bounded above, so has a supremum, say x_0. For $m,n \geq N$, we have $x_n \leq x_m + \varepsilon e$, so $x_0 \leq x_m + \varepsilon e$. Hence $\|x_m - x_0\| \leq \varepsilon$ $(m \geq N)$.

3.8. Cones with bases

Let P be a cone in a linear space X, and let B be a base for P. We recall from section 1.9 that there is a linear functional (unique if P generates X) such that

$$B = \{x \in P : f(x) = 1\}.$$

If f is continuous, then the following two facts are obvious:

(i) $0 \notin \bar{B}$

(ii) If P is closed, then so is B.

Conversely, if P gives an open decomposition of X and $0 \notin \bar{B}$, then f is continuous.

3.8.1. Let X be a complete, metrizable topological linear space, and let P be a closed cone generating X. Then any base for P is closed.

Proof. Any base for P is of the form $\{x \in P : f(x) = 1\}$ for some linear functional f. By 3.5.6, f is continuous.

One of our standard examples shows that 3.8.1 fails in the absence of completeness. Consider the space F with the supremum norm. A base for the cone P of non-negative sequences in F is $B = \{x \in P : \Sigma \xi_n = 1\}$, which is not closed, since $x_n \in B$, where

$$x_n = n^{-1} (e_1 + \dots + e_n),$$

and $x_n \to 0$.

Well-based cones

We say that a cone P in a topological linear space is well-based if it has a bounded base B such that $0 \notin \bar{B}$. Notice that this property is invariant under all the topologies of a dual pair.

Let B be a base as above. We recall that, by 1.9.4, the sets

$$C = \cup\{\lambda B : 0 \leq \lambda \leq 1\}, \qquad C' = \cup\{\lambda B : \lambda > 1\}$$

are convex. Since $0 \notin \bar{B}$, there exists circled U in $\mathfrak{N}(X)$ such that $U \cap B = \emptyset$. If $b \in B$ and $\lambda b \in U$, then $\lambda < 1$, so $P \cap U \subseteq C$.

3.8.2. A well-based cone is self-allied.

Proof. Let B be a bounded base such that $0 \notin \overline{B}$, and let U be a circled neighbourhood of 0 disjoint from B. Given circled V in $\mathfrak{N}(X)$, there exists $\alpha > 0$ such that $\alpha B \subseteq V$. If $0 \leq y \leq x \in \alpha U$, then there exist a,b in B and $\lambda, \mu \geq 0$ such that $x = \lambda a$, $y = \mu b$. Then $\lambda < \alpha$, and $\mu \leq \lambda$, by 1.9.2. Hence $y \in \alpha B \subseteq V$.

Example. We give a simple direct proof that the usual positive cones in m and c_o have no bounded bases, though this follows from results below. If B is a base contained in the unit ball, then, for each n, there exists λ_n in $(0,1]$ such that $\lambda_n e_n \in B$. Let

$$x_n = n^{-1}(\lambda_1 e_1 + \dots + \lambda_n e_n).$$

By convexity, $x_n \in B$. But $x_n \to 0$, and B is closed, by 3.8.1. This is a contradiction.

3.8.3. If X is a Hausdorff topological linear space and P is a cone in X having a bounded, closed base B, then P is closed.

Proof. Let $C = \bigcup\{\lambda B : 0 \leq \lambda \leq 1$. We show that C is closed. We can then deduce the result as follows. There is a closed neighbourhood W of 0 such that $P \cap W \subseteq C$. Then $P \cap W = C \cap W$, which is closed. The result follows, by 3.1.1.

Take x not in C. For each λ in $[0,1]$, there exists $U(\lambda)$ in $\mathfrak{N}(X)$ such that $x + U(\lambda)$ is disjoint from λB. Take circled $V(\lambda)$ in $\mathfrak{N}(X)$ such that $V(\lambda) + V(\lambda) \subseteq U(\lambda)$. There exists $\delta(\lambda) > 0$ such that $\alpha B \subseteq V(\lambda)$ for $|\alpha| \leq \delta(\lambda)$. Then $[x + V(\lambda)] \cap (\mu B) = \emptyset$ for all μ such that $|\mu - \lambda| \leq \delta(\lambda)$. There exist $\lambda_1, \dots, \lambda_n$ such that

$$[0, 1] \subseteq \bigcup_{i=1}^{n} [\lambda_i - \delta(\lambda_i), \ \lambda_i + \delta(\lambda_i)].$$

Let $V = \bigcap_{i=1}^{n} V(\lambda_i)$. Then $(x + V) \cap C = \emptyset$.

Returning to the correspondence between bases and strictly monotonic linear functionals, we obtain a duality theory between well-based cones and wedges with interior points. In the following, polars (and in particular, P^o) will be evaluated in the space X^* of continuous linear functionals on X.

<u>3.8.4. THEOREM</u>. Let P be a cone in a locally convex space X. Then P is well-based iff P^o has an interior point with respect to the strong topology $\beta(X)$ for X^*.

<u>Proof</u>. (i) Suppose that B is a bounded base for P such that $0 \notin \overline{B}$. Then there exists f in X^* such that $f \geq 1$ on B. Now B^o is a neighbourhood of 0 in the topology $\beta(X)$ for X^*, and for g in B^o, $f + g \geq 0$ on B, so $f + g \in P^o$.

(ii) Suppose that $f \in P^o$ and that V is a $\beta(X)$-neighbourhood of 0 such that $f + V \subseteq P^o$. Let $B = \{x \in P : f(x) = 1\}$. Clearly, B is a base for P, and $0 \notin \overline{B}$. Also, V^o is a bounded subset of X, and for x in B and g in V, we have

$$(f + g)(x) = 1 + g(x) \geq 0,$$

so $B \subseteq V^o$.

<u>3.8.5. COROLLARY</u>. Let P be a cone in a locally convex, Hausdorff space X. If P is well-based, then so is $\overline{P}$. If P is closed and well-based, then it has a bounded, closed base.

<u>Proof</u>. The proof of 3.8.4 shows that if P is well-based, then it has a bounded base of the form $B = \{x \in P : f(x) = 1\}$, where $f \in X^*$. We show that $\overline{B}$ is a base for $\overline{P}$, proving both statements. Now $\overline{B}$ is bounded and contained in $\{x \in \overline{P} : f(x) = 1\}$. Hence pos $\overline{B}$ is closed, by 3.8.3. It follows that pos $\overline{B} = \overline{P}$.

Instead of reasoning as in 3.8.5, it is tempting simply to take a bounded base B such that $0 \notin \overline{B}$, and to consider its closure. However, it is not clear that this closure still has the property that each line through 0 meets it at most once.

3.8.6. THEOREM. Let P be a wedge in a locally convex space X. Then:

(i) If P has an interior point, then P^o has a $\sigma(X)$-compact base.

(ii) If X is a Mackey space, P is closed and P^o has a $\sigma(X)$-compact base, then P has an interior point.

Proof. (i) Let e be an interior point of P, and let

$$B = \{f \in P^o : f(e) = 1\}.$$

Then B is a base for P^o. Now $(P - e) \in \mathfrak{N}(X)$, so $(P - e)^o$ is $\sigma(X)$-compact. The result follows, since B is a $\sigma(X)$-closed subset of $(P - e)^o$.

(ii) Suppose that B is a $\sigma(X)$-compact base for P^o. Then there is an element x of X such that $f(x) \geq 1$ for f in B. Since X is a Mackey space, $B^o \in \mathfrak{N}(X)$. For y in B^o and f in B, $f(x + y) \geq 0$, so $x + y \in P^{oo} = P$. Hence $x + B^o \subseteq P$.

The following example shows that P^o can be well-based with respect to $\beta(X)$ when P has no interior point. Let X be c_o with the usual positive cone P. Then P has no interior point, and P^o is the usual positive cone in l_1. The set of sequences $\{\xi_n\}$ in P^o such that $\Sigma\xi_n = 1$ is a base for P^o that is bounded and closed in the norm topology.

Increasing nets and suprema

By a residual subset of a directed net $\{x_n : n \in D\}$ we mean a set of the form $\{x_n : n \geq n_o\}$ for some n_o in D.

3.8.7. THEOREM. If the positive cone in an ordered topological linear space is well-based, then each increasing directed net in the space having a bounded residual subset is Cauchy.

Proof. Let B be a bounded base for the positive cone such that $0 \notin \bar{B}$. Given circled U in $\mathfrak{N}(X)$, take $\varepsilon > 0$ such that $\varepsilon B \subseteq U$. Suppose that $\{x_n : n \geq n_o\}$ is a bounded residual subset of the given net, and let $y_n = x_n - x_{n_o}$ $(n \geq n_o)$. For each $n \geq n_o$, there exist b_n in B and

$\lambda_n \geq 0$ such that $y_n = \lambda_n b_n$. Since $\{y_n : n \geq n_o\}$ is bounded, it is contained in a scalar multiple of $\cup\{\lambda B : 0 \leq \lambda \leq 1\}$, so $\{\lambda_n : n \geq n_o\}$ is bounded above. Let $\alpha = \sup \{\lambda_n : n \geq n_o\}$, and take $n_1 \geq n_o$ such that $\lambda_{n_1} \geq \alpha - \varepsilon$. For $n \geq n_1$, we have $y_n - y_{n_1} = \varepsilon_n c_n$, where $c_n \in B$ and $\varepsilon_n = \lambda_n - \lambda_{n_1} \leq \varepsilon$, by 1.9.2. Hence $x_n - x_{n_1} \in U$ $(n \geq n_1)$.

<u>3.8.8. COROLLARY</u>. Let X be a complete topological linear space, ordered by a closed, well-based cone. Then each majorised, increasing directed net in X converges to its supremum.

<u>Proof</u>. Let $\{x_n\}$ be an increasing net, bounded above by x_o. For each n_o, $\{x_n : n \geq n_o\}$ is contained in the order-interval $[x_{n_o}, x_o]$, so is bounded, by 3.8.2 and 3.2.6. By 3.8.7, the net is Cauchy, so has a limit. By 3.1.14, the limit is the supremum of the points in that net.

The next corollary is a conditional converse to 3.7.4. Let us say (temporarily) that an ordered set is <u>quasi-order-complete</u> if each majorised, upward-directed subset has a supremum. For lattice orderings, this obviously implies order-completeness.

<u>3.8.9. COROLLARY</u>. Let X be a topological linear space, ordered by a closed, well-based cone. If X is complete, then X is quasi-order-complete.

<u>Proof</u>. This follows from 3.8.8 on regarding an upward-directed set as a net directed by the given ordering.

Consider the following condition (which we refer to as (1)):

If A is an upward-directed set having a supremum a_o, then A (regarded as a net directed by the ordering) converges to a_o.

If the space is locally order-convex, then 3.2.9 shows that it is sufficient if, under the hypotheses of (1), a_o is in the closure of A. Heavy use of conditions related to (1) is made in the Japanese literature on linear lattices with norm topologies.

3.8.8 shows that (1) holds in a complete space with an ordering given by a closed, well-based cone. The example following 3.1.14 shows that (1) does not hold in the space m. It is not hard to verify that (1) does hold in c_o (though, as we have seen, the positive cone in c_o is not well-based).

A further corollary of 3.8.7 is a vital step in the proof of the Bishop-Phelps theorem (1) on support points.

3.8.10. COROLLARY. If X is a topological linear space ordered by a closed, well-based cone, and A is a bounded, complete subset of X, then each element of A precedes a maximal element of A.

Proof. Let B be a totally ordered subset of A. Regarded as a net directed by the ordering of X, B is Cauchy, by 3.8.7. Hence B has a limit in A, which is the supremum of B, by 3.1.14. The result follows, by Zorn's lemma.

Normed spaces

It is interesting to compare the next result with 3.7.1.

3.8.11. If X is a topological linear space, and there is a well-based cone that gives an open decomposition of X, then the topology of X is seminormable.

Proof. Let B be a bounded base for the cone such that $0 \notin \overline{B}$, and let $C = \bigcup\{\lambda B : 0 \leq \lambda \leq 1\}$. Then C - C is a bounded, convex neighbourhood of 0.

We say that a linear functional f on an ordered normed linear space X is uniformly monotonic if there exists $\alpha > 0$ such that $f(x) \geq \alpha \|x\|$ for all $x \geq 0$.

3.8.12. Let X be an ordered normed linear space with positive wedge P. Then an element f of X^* is uniformly monotonic iff it is an interior point of P^o with respect to the norm topology of X^*. Consequently, P is well-based iff there is a uniformly monotonic, continuous linear functional on X.

<u>Proof</u>. (i) Suppose that $\alpha > 0$ and $f(x) \geq \alpha\|x\|$ $(x \in P)$. Take g in X^* with $\|g\| < \alpha$ Then, clearly, $f + g \in P^o$.

(ii) Suppose that $f \in X^*$ and $f + g \in P^o$ whenever $g \in X^*$ and $\|g\| < \alpha$. Take x in P. There exists g in X^* such that $\|g\| = \alpha$ and $g(x) = \alpha\|x\|$. Then $(f - g)(x) \geq 0$, so $f(x) > \alpha\|x\|$.

<u>3.8.13</u>. Suppose that P is a cone in a normed linear space X, and that there exists $\alpha > 0$ such that, for x,y in P,

$$\|x + y\| \geq \|x\| + \alpha\|y\|.$$

Then P is well-based.

<u>Proof</u>. Let $C = \{x \in P : \|x\| > \alpha^{-1}\}$. We show that co(C) is disjoint from the closed unit ball U, so that there is a linear functional f in U^o such that $f(x) > 1$ for x in C. Then $f(x) \geq \alpha\|x\|$ for x in P.

Suppose that all convex combinations of n - 1 elements of C are out of U. Then a convex combination of n elements of U can be expressed in the form $\lambda x + (1 - \lambda)y$, where $0 < \lambda < 1$, $\|x\| > 1$ and $\|y\| > \alpha^{-1}$. Then

$$\begin{aligned}\|\lambda x + (1 - \lambda)y\| &\geq \lambda\|x\| + \alpha(1 - \lambda)\|y\| \\ &> \lambda + (1 - \lambda) = 1.\end{aligned}$$

This applies for each $n \geq 2$, and the result follows, by induction.

A simple example shows that the converse of 3.8.13 is not true. Consider R^2 with the usual order and the supremum norm. A uniformly monotonic linear functional is defined by $f(\xi,\eta) = \xi + \eta$, but each of (1,0), (0,1), (1,1) has norm one.

We prove later (4.4.4) that the positive cone for a lattice ordering can only be well-based and have interior points if the space is finite dimensional. If we are prepared to do without a lattice ordering, however, there is no shortage of well-based cones with interior points. A natural way of constructing such cones in a normed space is as follows: take f in X* with $\|f\| = 1$, take α in $(0,1)$, and let

$$K(f,\alpha) = \{x : f(x) \geq \alpha\|x\|\}.$$

Obviously, f is uniformly monotonic with respect to this cone. To show that $K(f,\alpha)$ has interior points, take $\varepsilon > 0$ such that $\alpha(1 + 2\varepsilon) < 1$. Then there exists x in X such that $\|x\| = 1$ and $f(x) > \alpha(1 + 2\varepsilon)$. If $\|y\| \leq \alpha\varepsilon$, then $\|x + y\| \leq 1 + \varepsilon$, and $f(x + y) \geq \alpha(1 + \varepsilon)$, so $x + y \in K(f, \alpha)$.

We finish this section by showing how this construction is used in proving the Bishop-Phelps theorem. If A is a subset of a topological linear space X, then a point x_0 of A is said to be a <u>support point</u> of A if there is a non-zero element f of X* such that $f(x_0) = \sup f(A)$.

<u>3.8.14. THEOREM</u>. If A is a complete, convex subset of a normed linear space X, then the support points of A are dense in the boundary of A.

<u>Proof</u>. Take $\varepsilon > 0$ and a boundary point x of A. Then there exists y not in A such that $\|y - x\| \leq \varepsilon/2$. Since A is closed, there exists f in X* such that $\|f\| = 1$ and $f(y) > \sup f(A)$. Let $K = \{x : f(x) \geq \frac{1}{2}\|x\|\}$, and let X have the ordering associated with K. Now $f(y - x) \leq \varepsilon/2$, so if $z \in A$ and $z \geq x$, then $f(z - x) \leq \varepsilon/2$, so $\|z - x\| \leq \varepsilon$. Hence $A \cap (x + K)$ is bounded, and 3.8.10 shows that there is a maximal element x_0 of $A \cap (x_0 + K)$. Then x_0 is clearly a maximal element of A, so $A \cap (x_0 + K) = \{x_0\}$, and $\|x_0 - x\| \leq \varepsilon$. Since K has interior points (and 0 is not one of them), there is a non-zero element g of X* such that $\inf g(x_0 + K) \geq \sup g(A)$. Then $g(x_0) = \sup g(A)$, so x_0 is a support point of A.

Ellis (4) has obtained some interesting results connecting support points with order structures and extreme points.

3.9. Base seminorms

Bearing in mind the result of 3.8.11, it is natural to define a seminorm corresponding to a base as follows. Let B be a base for a cone P in a linear space X. The convex cover of $B \cup (-B)$ is

$$\Delta(B) = \{\lambda a - (1-\lambda)b : a,b \in B \; ; \; 0 \leq \lambda \leq 1\}.$$

If $P - P = X$, then $\Delta(B)$ is absorbent, so we can define its Minkowski functional. This is a seminorm, called the <u>base seminorm</u> corresponding to B.

Since $\Delta(B)$ contains 0, it contains all elements of the form $\lambda a - \mu b$, where $a,b \in B$, $\lambda,\mu \geq 0$ and $\lambda + \mu \leq 1$.

If $C = \bigcup\{\lambda B : 0 \leq \lambda \leq 1\}$, then $\frac{1}{2}(C - C) \subseteq \Delta(B) \subseteq C - C$, so the Minkowski functional of $C - C$ is an equivalent seminorm. It will become apparent below why $\Delta(B)$ is preferred to $C - C$ in defining the base seminorm.

The conditions in 3.8.11 can be considered separately, as follows. Let τ be some topology for X, and let τ_B be the topology given by the seminorm corresponding to a base B. If B is bounded with respect to τ, then so is $\Delta(B)$, so $\tau \subseteq \tau_B$. If P gives an open decomposition of X with respect to τ, and 0 is not in the τ-closure of B, then $C - C$ is a τ-neighbourhood of 0, so $\tau_B \subseteq \tau$.

<u>3.9.1</u>. Let $\| \ \|$ be the seminorm corresponding to the base B. Then:

(i) $B = \{x \in P : \|x\| = 1\}$.

(ii) There exists f in X* such that $\|f\| = 1$ and $f(x) = \|x\|$ for x in P.

(iii) P is α-generating for all $\alpha > 1$.

(iv) If f is a linear mapping on X into a normed linear space Y, then $\|f\| = \sup\{\|f(b)\| : b \in B\}$.

<u>Proof</u>. (i) Suppose that $x \in B$, $\alpha > 0$ and $\alpha^{-1}x \in \Delta(B)$. Then there exist a,b in B and λ in $[0,1]$ such that $x = \alpha(\lambda a - \lambda' b)$, where $\lambda' = 1 - \lambda$. By 1.9.2, $\alpha(\lambda - \lambda') = 1$, so $\alpha \geq 1$. Hence $\|x\| = 1$.

If $x \in P$ and $\|x\| = 1$, then there exist b in B and $\lambda > 0$ such that

$x = \lambda b$. By the result just proved, $\lambda = 1$, so $x \in B$.

(ii) Let f be the linear functional taking the value 1 on B (it is unique, since $P - P = X$). Clearly, $|f(x)| \leqslant 1$ for x in $\Delta(B)$, so $\|f\| = 1$.

(iii) If $\|x\| < 1$, then $x \in \Delta(B)$. The result follows from (i).

(iv) This follows from 3.6.1 and (iii).

Part (i) of 3.9.1 shows that a base seminorm makes the cone well-based. It is easily seen that it also makes the cone 3-normal.

Examples. (i) l_1, usual order. The set of non-negative sequences $\{\xi_n\}$ such that $\Sigma\xi_n = 1$ is a base. We show that the corresponding base seminorm p is equal to the usual norm $\| \ \|$. By 3.9.1(i), $p(x) = \|x\|$ for $x \geqslant 0$. If $\|x\| \leqslant 1$, then there exist y, z in P such that $x = y - z$ and $\|y\| + \|z\| \leqslant 1$. Hence $p(x) \leqslant p(y) + p(z) \leqslant 1$. If $p(x) < 1$, then 3.9.1(iii) shows that there exist y', z' in P such that $x = y' - z'$ and $p(y') + p(z') \leqslant 1$ Hence $\|x\| \leqslant 1$.

(ii) In contrast to the situation for order-units, two different bases may give non-equivalent seminorms. Consider m with the usual ordering. Two bases are:

$$\{x \in P : \sum_{n=1} 2^{-n} \xi_n = 1\}, \quad \{x \in P : \sum_{n=1} 3^{-n} \xi_n = 1\}.$$

The corresponding seminorms (they are, in fact, norms, as we shall see in 4.4) take the values 2^{-n} and 3^{-n} respectively at e_n.

The duality of order-unit norms and base norms

The following results, due to Edwards and Ellis, are essentially numerical forms of 3.8.4 and 3.8.6.

3.9.2. Let X be an ordered normed linear space with positive wedge P, and suppose that the norm of X is that corresponding to an order-unit e. Suppose that Y is a linear subspace of X^* with the property that

$P^o \cap Y$ is α-generating in Y for all $\alpha > 1$. Then the norm of Y is the base norm corresponding to B, where $B = \{f \in P^o \cap Y : f(e) = 1\}$.

Proof. For f in P^o, $\|f\| = f(e)$, by 3.7.2, so $\|f\| = 1$ for f in B. Hence $\|f\| \leq 1$ for f in $\Delta(B)$.

Suppose that $f \in Y$ and $\|f\| < 1$. Then there exist g,h in $P^o \cap Y$ such that $f = g - h$ and $\|g\| + \|h\| < 1$. If g or h is zero, then it is clear that $f \in \Delta(B)$. If g and h are non-zero, then $g', h' \in B$, where $g' = g/g(e)$, $h' = h/h(e)$. Hence, in any case, $f \in \Delta(B)$.

3.9.3. COROLLARY. (i) If X has an order-unit norm, then X* has a base norm, given by a $\sigma(X)$-compact base.

(ii) If X is a Banach space with a closed cone, and X* has an order-unit norm, then X has a base norm.

Proof. (i) The cone P in X is 1-normal, by 3.7.2, so P^o is 1-generating in X*, by 3.6.2.

(ii) The cone in X is α-generating for all $\alpha > 1$, by 3.6.5.

3.9.4. Let X be an ordered normed linear space, the norm being that corresponding to a base B for the positive cone. Let h be the linear functional taking the value 1 on B, and let Y be a linear subspace of X* that contains h. Then the norm of Y is the order-unit norm corresponding to h.

Proof. For any f in Y, $|f(b)| \leq \|f\|$ $(b \in B)$, so

$$-\|f\|h \leq f \leq \|f\|h.$$

Hence h is an order-unit in Y, and $p_h(f) \leq \|f\|$ $(f \in Y)$, where p_h is the corresponding order-unit norm. If $-\lambda h \leq f \leq \lambda h$, then $|f(b)| \leq \lambda$ for b in B, so $\|f\| \leq \lambda$, by 3.9.1(iv). Hence $p_h(f) = \|f\|$.

3.9.5. COROLLARY. (i) If X has a base norm, then X* has an order-unit norm.

(ii) If X is a Banach space with a closed cone, and X* has a base norm, given by a $\sigma(X)$-compact base, then X has an order-unit norm.

Proof. (i) Clear

(ii) Let h be the linear functional on X* taking the value 1 on B. To show that h is in X, it is sufficient, by the Krein-Šmul'jan theorem (0.3.9), to show that the set

$$\{f \in X^* : \|f\| \leq 1 \text{ and } h(f) = 0\}$$

is $\sigma(X)$-closed. But this set is equal to $\frac{1}{2}(B - B)$, and so is $\sigma(X)$-compact.

The usual norm of c_o is not an order-unit norm (taking the usual order), although its dual, l_1, has a base norm (cf. the remarks following 3.8.6). Ng (2) has characterised the Banach spaces whose duals have base norms as those with "approximate order-unit norms" (and, equivalently, as those whose open unit balls are order-convex and directed).

Ellis (3) has shown how to generalise some of the above results to spaces of linear mappings.

3.10 Intrinsic topologies

If X is an ordered set, then the ordering of X can be used to define various topologies on X, particularly when X is a lattice (see Birkhoff (1), Roberts (1)). When X is a linear space, however, these topologies are not usually compatible with the linear structure. In this section, we show how two intrinsic locally convex topologies can be defined in an ordered linear space. The first of these agrees with the usual topology in all of our standard "natural" examples. Our treatment is largely directed towards the problem of recognising these topologies in particular cases.

The order-bound topology

The order-bound topology (or "order topology") τ_b for an ordered linear space $(X, \leq)$ is the Ⓐ-absorbing topology, where Ⓐ denotes the family of all order-intervals. In other words, a convex set U is a

τ_b-neighbourhood of 0 iff it absorbs each order-interval.

The order-bound topology is the largest locally convex topology making all order-intervals bounded. For if τ is a locally convex topology with this property, then each τ-neighbourhood of 0 absorbs all order-intervals, so is a τ_b-neighbourhood of 0. In particular, if τ is locally order-convex, then $\tau \subseteq \tau_b$.

__3.10.1__. If U is a convex, absorbent set that absorbs each order-interval of the form $[0,a]$, then U is a τ_b-neighbourhood of 0.

__Proof__. Take an order-interval $[a,b]$. There exists $\delta > 0$ such that for $0 < \lambda < \delta$, $\lambda a \in U$ and $\lambda[0, b-a] \subseteq U$. For such λ, we have $\frac{1}{2}\lambda[a,b] = \frac{1}{2}\lambda a + \frac{1}{2}\lambda[0, b-a] \subseteq U$.

__3.10.2__. If the positive wedge P generates X, then it gives an open decomposition of X with respect to τ_b.

__Proof__. Let U be a convex τ_b-neighbourhood of 0, and let $V = P \cap U - P \cap U$. Then V is convex and absorbent, and absorbs each order-interval of the form $[0, a]$.

__3.10.3__. If X is an ordered linear space in which order-units exist, then the order-bound topology for X is the topology induced by each of its order-unit seminorms.

__Proof__. Let e be an order-unit. Then $[-e,e]$ is convex and absorbs order-intervals, so is a τ_b-neighbourhood of 0. Conversely, any τ_b-neighbourhood of 0 absorbs $[-e,e]$. Hence the scalar multiples of $[-e,e]$ form a local base for τ_b.

Hence the order-bound topology for m or $C[0,1]$ (with the usual order) is the usual topology.

Since base-seminorm topologies are locally order-convex, they are not larger than τ_b. We have seen that different bases can give non-equivalent seminorms, so a base-seminorm topology can be strictly smaller than τ_b.

An order-unit seminorm is a norm iff the ordering is almost Archimedean. The implication one way can be generalised:

3.10.4. If the order-bound topology is Hausdorff, then the ordering is almost Archimedean.

Proof. If $-y \leq nx \leq y$ for all n in ω, then x is clearly in every τ_b-neighbourhood of 0.

3.10.5. Let X be an ordered linear space, Y a locally convex space, and f a linear mapping from X to Y. Then f is continuous with respect to the order-bound topology of X and the topology of Y iff it maps order-intervals into bounded sets.

Proof. f is continuous iff, given convex V in $\mathfrak{N}(Y)$ and an order-interval $[x_1, x_2]$ in X, $f^{-1}(V)$ absorbs $[x_1, x_2]$. This is equivalent to saying that V absorbs $f[x_1, x_2]$.

In particular, the set of τ_b-continuous linear functionals on X is precisely the set X^b of order-bounded functionals, so τ_b is Hausdorff iff X^b separates points of X. If τ is a locally convex topology such that all τ-continuous linear functionals are order-bounded, then all order-intervals are τ-bounded, so $\tau \subseteq \tau_b$. Hence τ_b is the Mackey topology induced on X by X^b. From the way in which τ_b was defined, it is immediate that it is bornological. Because of this, if it is also sequentially complete, then it is barrelled (see, e.g., Kelley-Namioka (1), 19.5). Some sufficient conditions for this to occur are given by Namioka (1) (4.10 and 8.5).

3.10.5 also shows that if X,Y are ordered linear spaces, then every order-bounded linear mapping from X to Y is continuous with respect to the order-bound topologies.

3.10.6. (Namioka). Suppose that X is an ordered linear space with a generating positive wedge P, and let τ_b denote the order-bound topology for X. If τ is a complete, metrizable topology for X with respect to which P is closed, then $\tau_b \subseteq \tau$.

Proof. Let d be an invariant metric inducing τ, and let $U_n = \{x : d(x,0) \leq 2^{-n}\}$. By 3.5.2, $P \cap U_n - P \cap U_n$ is a τ-neighbourhood of 0. For each n, $P \cap U_n - P \cap U_n$ is not contained in 2nV, so there exists x_n in $P \cap U_n \sim nV$. Then $\sum_{n=1}^{\infty} x_n$ converges (with respect to τ) to an element x of P, and $0 \leq x_n \leq x$ for each n, so V does not absorb $[0,x]$. Hence V is not a τ_b-neighbourhood of 0.

Notice that 3.5.5 follows at once from 3.10.5 and 3.10.6.

If the conditions of 3.10.6 hold and τ is also locally order-convex, then $\tau_b = \tau$. It follows that the order-bound topologies corresponding to the usual orderings coincide with the usual topologies of s, c_0 and l_1.

A rather different example is provided by the space F with the usual ordering (since there is no "usual topology" here, τ_b can't coincide with it!). All linear functionals on F are order-bounded, being represented by sequences in the usual way. It follows from the remarks after 3.10.5 that τ_b is the largest locally convex topology for F. (This topology is quite easy to describe: a local base is the family of sets of the form

$$\{x \in F : \Sigma\, \alpha_n |\xi_n| \leq 1\}$$

where $\{\alpha_n\}$ is a sequence of positive numbers.)

By 3.2.2, if τ_b is locally order-convex, then every order-bounded linear functional is the difference between two monotonic ones. So the space defined in 1.7 for which this property fails has an order-bound topology that is not locally order-convex. However, 3.5.9 shows that if the order-bound topology is complete and metrizable and makes P closed, then it is locally order-convex. We shall see in section 4.2 that this is also the case in linear lattices.

The topology of uniform convergence on order-intervals

Let (X,Y) be a dual pair of linear spaces, and let P be a wedge in X. In the following, P^o will always be evaluated in Y, and X,Y will be supposed to have the orderings associated with P,P^o respectively. If P^o generates Y, then every order-interval in Y is contained in one of the form $[-f,f]$. We suppose also that each element of X, when regarded as a linear functional on Y, is order-bounded (a sufficient condition for this is that P should generate X). The sets $[-f,f]^o$ $(f \in P^o)$ are then absorbent in X, and form a local base for a topology on X, which we denote by $\pi(Y)$. This topology has been studied by Peressini (1),(2); here we restrict ourselves to quite elementary properties.

The construction ensures that each element of P^o, and therefore each element of Y, is continuous with respect to $\pi(Y)$. Recognition of the dual space is completed by the following:

3.10.7. If Y is order-convex in X^b, then Y is the space of $\pi(Y)$-continuous linear functionals on X.

Proof. Let f be a $\pi(Y)$-continuous linear functional on X. Then there exists g in P^o such that $|f(x)| \leq 1$ for all x in $[-g,g]^o$. If $x \in P$ and $g(x) \leq 1$, then $x \in [-g,g]^o$, so $\pm f(x) \leq 1$. Hence $-g \leq f \leq g$, so $f \in Y$.

3.10.8. The topology $\pi(Y)$ is locally order-convex.

Proof. If $x \in [-f,f]^o$ and $0 \leq y \leq x$, then $y \in [-f,f]^o$, by 1.5.4. The result follows, by 3.2.3.

3.10.9. If τ is a topology with respect to which all elements of Y are continuous, and if P gives an open decomposition of X with respect to τ, then $\pi(Y) \subseteq \tau$.

Proof. Take f in P^o. Let $U = \{x \in X: f(x) < 1\}$, and write $P \cap U - P \cap U = V$. Then V is a τ-neighbourhood of 0. If $u \in P \cap U$ and $-f \leq g \leq f$, then $|g(u)| \leq 1$. It follows that $|g(v)| \leq 2$ for v in V. Hence $[-f,f]^o$ contains $\frac{1}{2}V$, so is a τ-neighbourhood of 0.

3.10.10. The topology $\pi(Y)$ is semi-normable iff Y contains an order-unit.

Proof. If h is an order-unit in Y, then it is clear that the scalar multiples of $[-h, h]^{o}$ form a local base for $\pi(Y)$. Conversely, if $\pi(Y)$ is semi-normable, then there is an element h of P^{o} such that $[-h,h]^{o}$ is bounded with respect to $\pi(Y)$. It follows with ease that $[-h,h]$ is absorbent in Y.

3.10.11. COROLLARY. If X^{b} contains an order-unit, then $\pi(X^{b}) = \tau_{b}$.

Proof. Since $\pi(X^{b})$ is a seminorm topology, it is the largest locally convex topology with dual space X^{b}.

If p is the Minkowski functional of $[-h,h]^{o}$, then it is easily seen that the dual norm to p is the order-unit norm corresponding to h.

Examples. (i) Let X be l_1 with the usual ordering. By 3.5.6 we know that X^{b} is isomorphic to m, which contains an order-unit. Hence $\pi(X^{b})$ is the usual norm topology.

(ii) Let X be c_o with the usual ordering. Then X^{b} is isomorphic to l_1, which contains no order-unit. By 3.9.9 and 3.10.10, it follows that $\pi(X^{b})$ is strictly smaller than the usual norm topology. We shall see in section 4.2 that it is larger than the weak topology $\sigma(l_1)$.

3.11. Nearly directed subspaces

It is elementary that if a linear subspace E of an ordered linear space X is directed, then E^{o} is order-convex in X'. The converse, however, is not true (see the example at the end of the section). In this section, we show how to characterise the subspaces E for which E^{o} is order-convex (in X', or, in the presence of a locally convex topology on X, in X*). The results are due to Ellis (2) and the author (3).

Remembering that a linear subspace E is directed iff, for each x in E, there exists y in E such that $y \geq x$ and $y \geq 0$, we define nearly directed subspaces as follows. Let E be a linear subspace of an ordered topological linear space X. Then E is said to be <u>nearly directed</u> (in X) if, given x in E and U in $\mathfrak{N}(X)$, there exist y in E and u_1, u_2 in U such that $y \geq x + u_1$ and $y \geq u_2$.

It is easily verified that the following statement is equivalent: given x in E and U in $\mathfrak{N}(X)$, there exist y in E and u_1, u_2 in U such that $-(y + u_2) \leq x \leq y + u_1$.

Clearly, the family of nearly directed subspaces becomes larger as the topology becomes smaller.

<u>3.11.1</u>. The closure of a nearly directed subspace is nearly directed.

<u>Proof</u>. Suppose that E is nearly directed. Take x in $\bar{E}$ and U in $\mathfrak{N}(X)$. There exists symmetric V in $\mathfrak{N}(X)$ such that $V + V \subseteq U$. Take v in V such that $x + v \in E$. There exist y in E and v_1, v_2 in V such that $y \geq x + v + v_1$ and $y \geq v_2$. The result follows.

<u>3.11.2</u>. X is nearly directed (in itself) iff P - P is dense in X.

<u>Proof</u>. Suppose that X is nearly directed. Take x in X and U in $\mathfrak{N}(X)$. There exists V in $\mathfrak{N}(X)$ such that $V - V \subseteq U$. There exist y in E, v_1, v_2 in V and p,q in P such that $x + v_1 = y - p$ and $v_2 = y - q$. Then $x + v_1 - v_2 = q - p$, so x + U meets P - P.

The converse is equally straightforward.

<u>3.11.3</u>. Suppose that the positive wedge P gives an open decomposition of X. Then a linear subspace E is nearly directed iff, given x in E and U in $\mathfrak{N}(X)$, there exist y in E and u in U such that $y + u \geq x$ and $y + u \geq 0$.

<u>Proof</u>. The condition is clearly sufficient. Suppose that E is nearly directed, and take x in E and U in $\mathfrak{N}(X)$. There exists U_1 in $\mathfrak{N}(X)$ such that $U_1 + U_1 \subseteq U$. Let $V = P \cap U_1 - P \cap U_1$. Then there exist y in E and v_1, v_2 in V such that $y \geq x + v_1$ and $y \geq v_2$. Now $v_i = u_i - u_i'$ for some u_i, u_i' in $P \cap U_1$ $(i = 1,2)$. Let $u = u_1' + u_2'$. Then $u \in U$ and $y + u \geq x$, $y + u \geq 0$.

We notice that everything that has been said so far can be said for commutative topological groups.

The main theorem

3.11.4 THEOREM. Let X be an ordered locally convex space, and let E be a linear subspace of X. Then E° is order-convex in X* iff E is nearly directed.

Proof. (i) Suppose that E is nearly directed, and take f,g in X* such that $f \in E^{\circ}$ and $0 \leq g \leq f$. Take x in E and $\varepsilon > 0$. There exists U in $\mathfrak{N}(X)$ such that $|f(u)| \leq \varepsilon$ and $|g(u)| \leq \varepsilon$ for u in U. There exist y in E and u_1, u_2 in U such that $y \geq x + u_1$ and $y \geq u_2$. Then

$$\begin{aligned} g(x) &\leq g(y - u_1) \\ &\leq g(y - u_2) + 2\varepsilon \\ &\leq f(y - u_2) + 2\varepsilon \\ &\leq 3\varepsilon. \end{aligned}$$

Hence $g(x) \leq 0$ for x in E, so $g \in E^{\circ}$.

(ii) Suppose that E is not nearly directed, and consider the space $X \times X$. There exist x_0 in E and U in $\mathfrak{N}(X)$ such that, given u_1, u_2 in U, y in E and p,q in P,

$$(x_0, 0) + (u_1, u_2) \neq (y-p, y-q).$$

Write

$$H = \{(y-p, y-q) : y \in E \text{ and } p,q \in P\}.$$

Then H is a wedge, and $(x_0, 0)$ is not in the closure of H. Hence there is a continuous linear functional φ on $X \times X$ such that $\varphi(x_0, 0) > \sup\varphi(H)$. Since H is a wedge, $\varphi \leq 0$ on H. Let $g(x) = \varphi(x,0)$, $h(x) = \varphi(0,x)$ $(x \in X)$. Then $g,h \in X^*$ and $\varphi(x,y) = g(x) + h(y)$. For p in P, $g(-p) = \varphi(-p,0) \leq 0$, so $g \in P^{\circ}$. Similarly, $h \in P^{\circ}$. For y in E, $g(y) + h(y) = \varphi(y,y) \leq 0$, so $g + h \in E^{\circ}$. But $g(x_0) > 0$, so $g \notin E^{\circ}$. Hence E° is not order-convex in X*.

3.11.5. COROLLARY. The same subspaces are nearly directed with respect to all the topologies of a dual pair.

3.11.6 COROLLARY. Let X be an ordered locally convex space with a closed positive wedge P. If E is a closed linear subspace of X, then E is order-convex iff E^o is nearly directed with respect to $\sigma(X)$.

Proof. This follows by applying 3.11.4 to the space X* with the topology $\sigma(X)$ and positive wedge P^o.

3.11.7. COROLLARY. Let X be an ordered locally convex space, and let P^o be the set of continuous, monotonic linear functionals on X. Then a non-zero element f of P^o is extremal in P^o iff its kernel is nearly directed.

Proof. f is extremal in P^o iff lin f is order-convex.

The characterisation of extremal elements of P^o given in 1.8.1 follows without difficulty (in fact, it is reasonable to regard 3.11.4 as a generalisation of 1.8.1).

3.11.8 COROLLARY. If E is nearly directed with respect to the order-bound topology, then it is nearly directed with respect to every locally convex topology for X.

Proof. Let X* be the space of linear functionals that are continuous with respect to a given locally convex topology, and take f,g in X* such that $f \in E^o$ and $0 \leq g \leq f$. Then $f,g \in X^b$, so $g \in E^o$.

Applying 3.11.4 to the largest locally convex topology, we obtain its purely algebraic form:

3.11.9. COROLLARY. Let E be a linear subspace of an ordered linear space X. Then the following statements are equivalent:

(i) E^o is order-convex in X'.

(ii) Given x in E and a convex, absorbent set U, there exist y in E and u_1,u_2 in U such that $y \geq x + u_1$ and $y \geq u_2$.

Perfect subspaces

Let X be an ordered linear space with an order-unit e. Bonsall (3) defined a subspace E of X to be perfect if, given x in E and $\varepsilon > 0$, there exists y in E such that $y + \varepsilon e \geq x$ and $y + \varepsilon e \geq 0$. It is easily seen that this definition is independent of the choice of order-unit, and that it implies that E is nearly directed with respect to any linear topology on X. Recalling that order-unit seminorms induce the order-bound topology, we can prove the following converse:

3.11.10. Let X be an ordered linear space with order-units, and let E be a linear subspace of X. Then E is perfect iff E is nearly directed with respect to the order-bound topology.

Proof. Suppose that E is nearly directed with respect to the order-bound topology. Let e be an order-unit, and let $\| \;\|$ be the corresponding seminorm. Given x in E and $\varepsilon > 0$, there exist y in E and u_1, u_2 such that $\|u_i\| < \varepsilon$ $(i = 1,2)$ and $y + u_1 \geq x$, $y + u_2 \geq 0$. Then $u_i \leq \varepsilon e$ $(i = 1,2)$, so $y + \varepsilon e \geq x$ and $y + \varepsilon e \geq 0$.

Kist (1) has defined "perfect" subspaces of spaces without order-units in such a way that perfect subspaces are nearly directed with respect to all linear topologies. However, it is not true that all linear subspaces that are nearly directed with respect to the order-bound topology are perfect in this sense, as is seen by considering a space whose order-bound topology is indiscrete (e.g. the space defined in section 1.5, ex. (v), that has no non-zero order-bounded linear functionals).

Example

Consider the space m with the usual norm and ordering. Define e and e_n as usual, and let a be the sequence having 1 in odd places and 0 in even places. Let $x_n = a - n^{-1} e_{2n}$, and let E be the subspace spanned by the x_n. Then $E \cap P = \{0\}$, so E is order-convex and not directed. We show that E is perfect (so nearly directed). A general element of E is

$$x = \sum_{i \in I} \lambda_i x_i - \sum_{j \in J} \mu_j x_j,$$

where each $\lambda_i, \mu_j > 0$ and I,J are disjoint, finite sets of integers. Let

$$\lambda = \sum_{i \in I} \lambda_i, \qquad \mu = \sum_{j \in J} \mu_j.$$

Given $\varepsilon > 0$, take $r > (\lambda \vee \mu)\varepsilon^{-1}$, and let

$$y = (\lambda \vee \mu)x_r - \sum_{j \in J} \mu_j x_j.$$

Then $y + \varepsilon e \geqslant x$ and $y + \varepsilon e \geqslant 0$.

Another interesting fact about this example is that $\bar{E}$ is not order-convex (although E is). For $a \in \bar{E}$, and $0 \leqslant e_1 \leqslant a$, but $e_1 \notin \bar{E}$, since sequences in $\bar{E}$ have all odd components equal.

3.12. Locally compact wedges and cones

In this section, more than anywhere else in this book, attention will be centered on wedges rather than the associated orderings. However, the results are connected with those of earlier sections of this chapter, especially 3.2, 3.8 and 3.9.

3.12.1. Let P be a wedge in a topological linear space X. Then P is locally compact iff $P \cap U$ is compact for some U in $\mathfrak{N}(X)$.

Proof. The condition is obviously necessary. Suppose that it holds. Given x in P, there exists $\lambda > 0$ such that x is an interior point of λU. Then $P \cap (\lambda U) = \lambda(P \cap U)$, and this set is a compact P-neighbourhood of x.

By 3.1.1, it follows that a locally compact wedge in a Hausdorff linear space must be closed. Clearly, if P is locally compact and gives an open decomposition of a Hausdorff linear space X, then X must be finite dimensional. Thus, by 3.5.2, an infinite dimensional complete, metrizable linear space is not generated by any locally compact wedge (though it may well be generated by a wedge that is locally compact in a weak topology, as we now see).

3.12.2. In a Hausdorff linear space, any cone with a compact base is locally compact.

Proof. Let B be a compact base for P. Then B is closed and does not contain 0, so there is a circled neighbourhood U of 0 disjoint from B. The set $\bigcup\{\lambda B : 0 \leq \lambda \leq 1\}$ is compact and contains $P \cap U$.

We shall see below (3.12.8) that the converse is true in locally convex spaces. If X is a locally convex space, and P is a wedge in X with non-empty interior, then 3.8.6 and 3.12.2 show that P^0 is locally compact with respect to $\sigma(X)$.

3.12.3. THEOREM. Let P,Q be wedges such that P is locally compact, Q is closed, and $P \cap (-Q) = \{0\}$. Then:

(i) P is allied to Q.

(ii) P + Q is closed.

(iii) If Q is locally compact, then so is P + Q.

Proof. (i) Any neighbourhood of 0 contains a circled neighbourhood U such that $P \cap U$ is compact. Let $P_1 = (P \cap U) \sim \text{int } U$. Then P_1 is compact, so $P_1 + Q$ is closed. Also, $0 \notin P_1 + Q$, so there exists circled V in $\mathfrak{N}(X)$ such that $V \cap (P_1 + Q) = \emptyset$. Take p in $P \sim U$ and q in Q. There exists λ in (0,1) such that $\lambda p \in P_1$. Then $\lambda(p + q) \notin V$, so $p + q \notin V$. Hence if $p \in P$, $q \in Q$ and $p + q \in V$, then $p \in U$.

(ii),(iii). There exists U in $\mathfrak{N}(X)$ such that $P \cap U$ is compact and $Q \cap U$ is closed in case (ii), compact in case (iii). By (i), there exists V in $\mathfrak{N}(X)$ such that

$$(P + Q) \cap V \subseteq P \cap U + Q \cap U.$$

The results follow, by 3.1.1 and 3.12.1.

<u>3.12.4 COROLLARY</u>. A locally compact cone in a Hausdorff linear space is self-allied.

<u>Example</u>

Let P be the usual positive cone in s. We show that pos e - P is not closed, so that we cannot dispense with the disjointness condition in 3.12.3. Let

$$x_n = ne - (n-1, n-2, \ldots, 1)$$
$$= (1,2, \ldots, n,n, \ldots).$$

Then

$$x_n \to (1,2, \ldots, n, n+1, \ldots),$$

which is not in pos e - P, since it is not bounded above.

<u>3.12.5</u>. Let X be a Hausdorff linear space with an ordering given by a locally compact cone P. If A is a compact subset of X, then [A] is compact. In particular, order-intervals are compact.

<u>Proof</u>. [A] is bounded, since P is self-allied. There exists U in $\mathfrak{N}(X)$ such that $P \cap U$ is compact. Take symmetric V in $\mathfrak{N}(X)$ such that $V + V \subseteq U$. There exists $\lambda > 0$ such that $[A] \subseteq \lambda V$. Take x in [A] Then there exist a in A and p in P such that $x = a + p$. Now a and x are in λV, so $p \in \lambda U$. Hence [A] is contained in $A + \lambda(P \cap U)$, which is compact. Since P is closed, [A] is closed, and the result follows.

The conclusion of 3.12.5 can hold when P is not locally compact, even in a complete, metrizable space (cf. 3.2.7). For 3.4.9 shows that the conclusion holds in s (with the usual topology and ordering), but the positive cone in s is not locally compact, since it gives an open decomposition of s.

The next result is a kind of open mapping theorem. The proof given here is due to Professor A.P. Robertson.

3.12.6. Let X,Y be Hausdorff linear spaces, and let P be a locally compact wedge in X. Let f be a continuous linear mapping from X to Y such that if $x \in P \sim \{0\}$, then $f(x) \neq 0$. Then:

(i) Given U in $\mathfrak{N}(X)$, there exists V in $\mathfrak{N}(Y)$ such that

$$f(P \cap U) \supseteq f(P) \cap V.$$

(ii) f(P) is locally compact.

Proof. (ii) is an immediate consequence of (i) and 3.12.1. To prove (i), take circled U in $\mathfrak{N}(X)$ such that $P \cap U$ is compact. Let $P_1 = (P \cap U) \sim \text{int } U$. Then P_1 is compact, so $f(P_1)$ is compact. Also, $0 \notin f(P_1)$, so there exists circled V in $\mathfrak{N}(Y)$ such that $V \cap f(P_1) = \emptyset$. Take x in $P \sim U$. There exists λ in (0,1) such that $\lambda x \in P_1$. Then $\lambda f(x) \notin V$, so $f(x) \notin V$. Thus if $x \in P$ and $f(x) \in V$, then $x \in U$.

Applying this result to the identity mapping, we obtain:

3.12.7 COROLLARY. Let τ_1, τ_2 be Hausdorff topologies for a linear space X, τ_2 being smaller than τ_1. If a wedge P in X is locally compact with respect to τ_1, then it is locally compact with respect to τ_2.

The rest of this section is concerned with locally convex spaces. The main result is:

3.12.8. THEOREM (Klee). If P is a cone in a locally convex, Hausdorff space, then the following statements are equivalent:

(i) P is locally compact;

(ii) there exists a continuous linear functional f such that $f \geq 0$ on P and $\{x \in P : f(x) \leq 1\}$ is compact;

(iii) P has a compact base.

Proof. (i) => (ii). Take convex U in $\mathfrak{N}(X)$ such that $P \cap U$ is compact. Let $P_1 = (P \cap U) \sim \text{int } U$. If $x \in P \sim \{0\}$, then $\lambda x \notin U$ for some $\lambda > 0$, since $P \cap U$ is compact, so there exists $\mu > 0$ such that $\mu x \in P_1$. Let $K = \overline{\text{co}}(P_1)$. Then K is a closed subset of $P \cap U$, so is compact. If $0 \in K$, then 0 is an extreme point of K, since P is a cone. By 0.3.7, this would imply that $0 \in P_1$, which is not so. Hence $0 \notin K$, so there exists f in X* such that $\inf f(K) \geq 1$. If $x \in P \sim U$, then $\lambda x \in P \sim U$ for $\lambda \geq 1$, so there exists λ in $(0,1)$ such that $\lambda x \in P_1$. Hence $f(x) > 1$. Therefore $\{x \in P : f(x) \leq 1\}$ is a closed subset of $P \cap U$, so is compact.

(ii) => (iii). Suppose that f is as in (ii). If $x \in P$ and $f(x) = 0$, then $x = 0$, since $\{\lambda x : \lambda > 0\}$ is contained in a compact set. Hence $\{x \in P : f(x) = 1\}$ is a compact base for P.

(iii) => (i). By 3.12.2.

As a first corollary, we have a form of the Krein-Mil'man theorem applicable to cones:

<u>3.12.9. COROLLARY</u>. A locally compact cone in a locally convex, Hausdorff space is the closed, convex cover of its extremal elements.

<u>Proof</u>. Let Q be the closed, convex cover of the extremal elements, and let B be a compact base. By 1.9.6, Q contains the extreme elements of B, so Q contains B. Since Q is clearly positive homogeneous, the result follows.

This result can be generalised as follows: define K to be a <u>cap</u> of P if K is compact and convex, and $P \sim K$ is convex. An extreme point of a cap is an extremal point of P, so if P is the union of its caps, then it is the closed, convex cover of its extremal points (see Phelps (1), pp. 87-89). Pryce (2) has obtained some interesting results on caps of convex sets.

We now give two more corollaries of 3.12.8.

<u>3.12.10 COROLLARY</u>. Let X be a locally compact, Hausdorff space, and let P be a locally compact cone in X. If Q is a closed wedge in X

such that $P \cap Q = \{0\}$, then there exists f in X^* such that

$$f(x) > 0 \quad (x \in P \sim \{0\}),$$
$$f(y) \leq 0 \quad (y \in Q).$$

Proof. P has a compact base B, which is disjoint from Q. Hence $B - Q$ is closed and does not contain 0, so there exists f in X^* such that $\inf f(B - Q) > 0$. Then $f > 0$ on B, and $f \leq 0$ on Q.

3.12.11 COROLLARY. Let P be a locally compact cone in a locally convex, Hausdorff space X. Suppose that F is a subset of X^* such that, given x in $P \sim \{0\}$, there exists f in F with $f(x) > 0$. Then there exists a finite subset $\{f_1, \ldots, f_n\}$ of F such that, given x in $P \sim \{0\}$, there exists i such that $f_i(x) > 0$.

Proof. P has a compact base B. The open sets $\{x : f(x) > 0\}$ $(f \in F)$ cover B, so a finite subfamily does.

Hustad (1) has proved the converse result that if P is closed and has the property expressed in 3.12.11, then P is locally compact with respect to $\sigma(X^*)$.

If P is locally compact and generates X, it is natural to ask what can be said about the base seminorm corresponding to a compact base of P. Some answers are given by the next result:

3.12.12. Let τ be a locally convex, Hausdorff topology for a linear space X, and let P be a cone that generates X and is locally compact with respect to τ. Let B be a τ-compact base for P. Then the base seminorm corresponding to B is a norm, with respect to which X is a Banach space. The norm topology is not smaller than τ. Any other τ-compact base for P gives an equivalent norm.

Proof. $\Delta(B)$ is τ-compact, since it is the image of $B \times B \times [0,1]$ under the continuous mapping $(a,b,\lambda) \to \lambda a - (1-\lambda)b$. It follows that the base seminorm is a norm, and that every τ-neighbourhood of 0 is a norm-neighbourhood. Since the unit ball is τ-complete, 0.3.8 shows that X is complete with respect to the norm. There exists a τ-neighbourhood U of 0 such that $P \cap U \subseteq B_1$, where $B_1 = \bigcup\{\lambda B : 0 \leq \lambda \leq 1\}$. So if C is another τ-compact base, then $\alpha C \subseteq B_1$ for some $\alpha > 0$. It follows with ease that the norms given by B and C are equivalent.

Edwards (1) has shown that the Banach space in 3.12.12 can always be expressed as the dual of another Banach space.

CHAPTER 4

LINEAR LATTICES WITH A TOPOLOGY

Let X be a linear lattice with a topology. The most natural connection to require between the topology and the lattice ordering is continuity of the lattice operations $x \to x^+$, $(x,y) \to x \vee y$, etc. These conditions are studied in section 4.1, separate consideration being given to the three cases in which $x \to x^+$ is (i) continuous at 0, (ii) continuous on X, (iii) uniformly continuous on X. The third of these conditions is the most interesting, and turns out to be equivalent to each of the following statements:

(a) P is self-allied and gives an open decomposition of X;

(b) the solid neighbourhoods of 0 form a local base.

This result, along with some others, holds equally for commutative lattice groups. In 4.2, we consider linear functionals, and examine how the conditions studied in 4.1 behave when the topology is varied. As in chapter 3, it is possible to give topological versions of some of the earlier algebraic results. Section 4.3 is devoted to spaces in which there is a local base consisting of sublattices. 3.1.12 tells us of the existence of plenty of continuous real lattice homomorphisms on such spaces, and we use this fact to represent such spaces as lattices of continuous functions. Lattices with order-unit norms belong to this class, and this yields a very natural approach to the Stone-Čech compactification. Lattices with base norms ("L-spaces") are studied briefly in section 4.4. The classical theorem on representation of L-spaces as spaces of measures is omitted, but we show how to derive some of the properties of L-spaces without having recourse to this theorem.

For further material (for example, results connecting order-completeness and completeness) we refer to the books of Peressini (2) and Fremlin (1). Among recent developments we mention the papers of Wong

(1,2,3). Ng (3) gives some generalisations to the case when the space is not a lattice.

4.1. Continuity of the lattice operations

Let X be a linear lattice with a topology. It is evident that if one of the mappings $x \to x^+$, $x \to x^-$, $x \to |x|$ is continuous at a point x_0, then so are the other two. We shall consider in turn spaces in which $x \to x^+$ is (i) continuous at 0, (ii) continuous on X, (iii) uniformly continuous on X.

Continuity at 0

If A is a subset of X, we write $A^+ = \{a^+ : a \in A\}$, and define A^- and $|A|$ similarly.

4.1.1. Let X be a linear lattice with a topology such that $x \to x^+$ is continuous at 0. Let P denote the positive wedge. Then:

(i) P gives an open decomposition of X.

(ii) If A is a bounded subset of X, then A^+, A^- and $|A|$ are bounded.

(iii) P gives a bounded decomposition of X.

Proof. (i) Given U in $\mathfrak{N}(X)$, there exists V in $\mathfrak{N}(X)$ such that if $x \in V$, then $x^+, x^- \in U$. Then $P \cap U - P \cap U \supseteq V$.

(ii) Given U in $\mathfrak{N}(X)$, take V as in (i). There exists $\lambda > 0$ such that $\lambda A \subseteq V$. Then $\lambda A^+ \subseteq U$.

(iii) follows from (ii).

We can argue conversely in metrizable spaces (cf. 3.2.7):

4.1.2. Let X be a linear lattice with a metrizable topology. Suppose that for each bounded subset A of X, A^+ is bounded. Then $x \to x^+$ is continuous at 0.

Proof. Let $\{U_n\}$ be a countable, contracting local base. If

$x \to x^+$ is not continuous at 0, then there exists U in $\mathfrak{N}(X)$ such that for each n, there exists x_n in $n^{-1}U_n$ with $x_n^+ \notin U$. Then $nx_n \to 0$, so the sequence $\{nx_n\}$ is bounded. However, $\{nx_n^+\}$ is unbounded.

<u>Example</u>. Consider R^2 with the lexicographic ordering. For each x, either $x^+ = x$ or $x^+ = 0$, so the mapping $x \to x^+$ is continuous at 0. It is discontinuous at $(0,-1)$, since $(n^{-1},-1)^+ = (n^{-1},-1)$, while $(0,-1)^+ = (0,0)$.

<u>Continuity on X</u>

The relation $x \vee y = (x - y)^+ + y$ shows that continuity of $x \to x^+$ on X is equivalent to continuity of $(x,y) \to x \vee y$ on $X \times X$ (where X is a commutative lattice group with a topology).

<u>4.1.3</u>. Let X be a linear lattice with a Hausdorff topology such that $x \to x^+$ is continuous on X. Let P denote the positive wedge. Then:

(i) P is closed.

(ii) Every band in X is closed.

<u>Proof</u>. (i) This follows from the fact that $P = \{x : x = x^+\}$.

(ii) The ordering is Archimedean, by (i), so if E is a band, then $E = E^{\perp\perp}$, by 2.4.7. But $A^\perp$ is closed for any subset A of X, since $x \to |x| \wedge |a|$ is continuous for each a in A.

Conversely, if P is closed, then the topology must be Hausdorff, by 3.1.2, since a lattice ordering is antisymmetric.

Recall the example at the end of 3.11 showing that, in an ordered topological linear space, the closure of an order-convex subspace need not be order-convex. For lattice orderings, there is a positive result. It is elementary that if X is a topological space with a lattice ordering such that the mappings $(x,y) \to x \vee y$ and $(x,y) \to x \wedge y$ are continuous, then the closure of a sublattice is a sublattice.

4.1.4. Let X be a commutative lattice group with a topology such that $x \to x^+$ is continuous on X. If E is a solid subgroup of X, then $\bar{E}$ is solid.

Proof. As just remarked, $\bar{E}$ is a sublattice. We show that $\bar{E}$ is also order-convex; the result then follows, by 2.3.10. Suppose that $x \in \bar{E}$ and $0 \leq y \leq x$. Given U in $\mathfrak{N}(X)$, there exists V in $\mathfrak{N}(X)$ such that for v in V, $(y + v)^+ - y \in U$. Take v in V such that $x + v \in E$. Then $0 \leq (y + v)^+ \leq (x + v)^+ \in E$, so $(y + v)^+ \in E$. Hence $y \in \bar{E}$.

Uniform continuity

We now come to what may be regarded as the basic theorem on topological linear spaces (or commutative groups) with lattice orderings. We say that such a space is locally solid if the solid neighbourhoods of 0 form a local base. (Of course, this does not imply that the solid neighbourhoods of any other point form a base of neighbourhoods of that point.)

4.1.5 THEOREM. Let X be a commutative lattice group with a topology, and let P denote the positive set. Then the following statements are equivalent:

(i) the mapping $(x,y) \to x \vee y$ is uniformly continuous on $X \times X$;

(ii) the mapping $x \to x^+$ is uniformly continuous on X;

(iii) P is self-allied and gives an open decomposition of X;

(iv) P is self-allied and $x \to x^+$ is continuous at 0;

(v) X is locally solid.

Proof. (i) => (ii). A priori.

(ii) => (iii). P gives an open decomposition of X, by 4.1.1. To show that P is self-allied, take U in $\mathfrak{N}(X)$. There exists V in $\mathfrak{N}(X)$ such that if $x - y \in V$, then $x^+ - y^+ \in U$. If $a,b \in P$ and $a + b \in V$, then $a = a^+ - (-b)^+ \in U$.

(iii) => (iv). Since P is self-allied, X is locally order-convex. Take order-convex U in $\mathfrak{N}(X)$, and let $V = P \cap U - P \cap U$. Then $V \in \mathfrak{N}(X)$. If $x \in V$, then there exist u_1, u_2 in $P \cap U$ such that $x = u_1 - u_2$. Then $0 \leq x^+ \leq u_1$, so $x^+ \in U$.

(iv) => (v). Take order-convex U in $\mathfrak{N}(X)$. There exists V in $\mathfrak{N}(X)$ such that if $x \in V$, then $\pm|x| \in U$. If $x \in V$ and $|y| \leq |x|$, then $-|x| \leq y \leq |x|$, so $y \in U$. Hence S(V), i.e.

$$\{y : |y| \leq |x| \quad \text{for some } x \text{ in } V\}$$

is a solid neighbourhood of 0 contained in U.

(v) => (i). Take solid U in $\mathfrak{N}(X)$. There exists solid V in $\mathfrak{N}(X)$ such that $V + V \subseteq U$. Take v_1, v_2 in V and x,y in X. Let

$$z = (x + v_1) \vee (y + v_2) - (x \vee y).$$

Then $v_1 \wedge v_2 \leq z \leq v_1 \vee v_2$, so $|z| \leq |v_1| + |v_2|$, and $z \in U$.

Statement (iii) enables us to relate uniform continuity of $x \to x^+$ to conditions which make sense when the ordering is not a lattice ordering. Roughly speaking, P must not be too large or too small.

We recall that a <u>lattice seminorm</u> is a seminorm p for which $|x| \leq |y|$ implies $p(x) \leq p(y)$.

<u>4.1.6</u>. If X is a locally solid, locally convex linear lattice, then the solid, convex neighbourhoods of 0 form a local base, and the topology can be given by lattice seminorms.

<u>Proof</u>. Take convex U in $\mathfrak{N}(X)$. There is a solid neighbourhood V contained in U. By 2.3.11, co(V) is solid, so is a solid, convex neighbourhood contained in U. It is obvious that the Minkowski functional of a solid, convex neighbourhood of 0 is a lattice seminorm.

The same reasoning shows that if the topology is normable and locally solid, then it can be given by a lattice norm. A linear lattice with a lattice norm will be called a <u>normed lattice</u>. Our "natural" examples m, c_o, l_1, C[0,1] are all normed lattices. In the terminology of the second duality theory of section 3.6, a normed lattice has properties (1) and (2) with constant 1 (but in the terminology of the first duality theory, we can only say that the cone is 2-normal and 2-generating).

4.1.7. A normed linear space with a lattice ordering is locally solid iff there exists $\delta > 0$ such that $|x| \geq |y|$ implies $\|x\| \geq \delta\|y\|$.

Proof. (i) If the space is locally solid, then there exists a lattice norm equivalent to the given norm. The stated condition follows with ease. (ii) If the condition holds, then the positive cone is self-allied, by 3.2.8, and $x \to |x|$ is continuous at 0.

4.1.8. If E is a subset of a locally solid commutative lattice group X, then E is allied to $E^\perp$. (Cf. 1.5.4).

Proof. Take solid U in $\mathfrak{N}(X)$. Suppose that $x \in E$, $y \in E^\perp$ and $x + y \in U$. By 2.4.2, $|x+y| = |x| + |y|$, so $x,y \in U$.

It follows that if E is a band and $E + E^\perp = X$ (as in an order-complete lattice), then the projection on E along $E^\perp$ is continuous.

Example

The following example shows that the mapping $x \to x^+$ can be continuous for all x without being uniformly continuous. Consider c_o with the usual norm and the ordering given by the "partial-sum cone" P_s. We saw in section 2.2 that this is a lattice ordering, and in section 3.2 that P_s is not self-allied. Consequently by 4.1.5, the mapping $x \to x^+$ is not uniformly continuous. We prove, however, that it is continuous at any point $x = \{\xi_n\}$ of c_o. Take $\varepsilon > 0$. There exists N such that $|\xi_r| \leq \varepsilon$ for $r > N$. Take $y = \{\eta_n\}$ in c_o such that $\|y - x\| \leq \varepsilon/N$. Let $X_r = \xi_1 + \ldots + \xi_r$ $(r \geq 1)$, $X_o = 0$, and define Y_r similarly. Then $x^+ = \{\alpha_n\}$, $y^+ = \{\beta_n\}$, where

$$\alpha_r = X_r^+ - X_{r-1}^+, \quad \beta_r = Y_r^+ - Y_{r-1}^+ \qquad (r \geq 1).$$

We use the fact that $|\lambda^+ - \mu^+| \leq |\lambda - \mu|$ for any real numbers λ, μ. For $r \leq N$, we have

$$|Y_r - X_r| \leq \sum_{s=1}^{r} |\eta_s - \xi_s| \leq \varepsilon,$$

so $|Y_r^+ - X_r^+| \leq \varepsilon$, and $|\beta_r - \alpha_r| \leq 2\varepsilon$. For $r > N$, $|\alpha_r| \leq |\xi_r| \leq \varepsilon$, and $|\beta_r| \leq |\eta_r| \leq \varepsilon + \|y - x\| \leq 2\varepsilon$, so $|\beta_r - \alpha_r| \leq 3\varepsilon$. Hence $\|y^+ - x^+\| \leq 3\varepsilon$.

By contrast, it is easily seen that in the space m with the P_s-ordering, $x \to x^+$ is discontinuous at $(1,-1,1,-1, \ldots)$. Roughly speaking, continuity of this mapping in c_o is due to the fact that each point is close to a finite dimensional subspace, but continuity is not uniform because the dimension of the subspace required depends on x.

4.2. Linear functionals, duality and different topologies

Confining attention to linear functionals, we can give topological versions of some of the results in 2.6. Starting with the Riesz decomposition theorem 2.6.1, we have:

4.2.1. Let X be a Riesz space with a topology such that the positive cone P is self-allied and gives an open decomposition of X. Let X^b denote the set of order-bounded linear functionals on X. Then X* is a solid subspace of X^b, so X* is an order-complete linear lattice, with the operations defined as in 2.6.1.

Proof. Since P is self-allied, $X^* \subseteq X^b$. Since P gives an open decomposition of X, X* is order-convex in X^b (3.3.3). Take f in X*. The supremum of f and 0 in X^b is f^+, where $f^+(x) = \sup f[0,x]$ $(x \in P)$. Take symmetric, order-convex U in $\mathfrak{N}(X)$ such that $\sup f(U) \leq 1$. If $x \in P \cap U$ and $0 \leq y \leq x$, then $y \in U$, so $f(y) \leq 1$. Hence $f^+(x) \leq 1$. Therefore $f^+ \leq 1$ on $P \cap U - P \cap U$, so $f^+ \in X^*$. Hence X* is a sublattice of X^b.

We recall that if P is self-allied, then either of the following conditions is sufficient for X* to be equal to X^b:

(i) P has an interior point (3.1.4),

(ii) X is complete and metrizable and P is closed (3.5.6).

In order to extend 4.2.1 to linear mappings with values in an order-complete lattice Y, we would need to assume that Y has a local base consisting of neighbourhoods admitting suprema. It will become clear in 4.3 that this is a very special condition.

For normed spaces, the second duality theory of 3.6 gives:

<u>4.2.2</u>. Let X be a normed Riesz space in which the following conditions hold:

(i) if $-x \leq y \leq x$, then $\|y\| \leq \|x\|$;

(ii) given x in X and $\varepsilon > 0$, there exists y in X such that $-y \leq x \leq y$ and $\|y\| \leq \|x\| (1 + \varepsilon)$.

Then X* is a normed lattice.

<u>Proof</u>. By 4.2.1, X* is a lattice. By 3.6.6 and 3.6.7, the cone X* has properties (N) and (G) with constant 1. It follows with ease that the norm in X* is a lattice norm.

By 3.6.7 and 3.6.8, we know that if X is a Banach space with a closed cone, and X* is a normed lattice, then conditions (i) and (ii) of 4.2.2 hold in X. It was shown by Andô (1) that, under these conditions, X is a Riesz space; other proofs have been given by Davies (1) and Ng(3).

The next result can be regarded as the topological form of 4.2.2, though we formulate it under the assumption that X is a lattice, not a Riesz space.

<u>4.2.3</u>. If X is a locally convex, locally solid linear lattice, then the strong topology β(X) for X* is locally solid.

Proof. A basic $\beta(X)$-neighbourhood of 0 is A^o, where A is a bounded, convex subset of X. Then the solid cover A' of A is bounded, since A is locally solid. Let $B = co(A')$. Then B is bounded and solid (2.3.11), so the polar of B in X^b is solid, by 2.6.2. It follows that the polar of B in X* is solid in X*. But this is a $\beta(X)$-neighbourhood of 0 contained in A^o.

If X is locally solid and barrelled, it can be shown that X* is a band in X^b, and that the strong topology for X* is complete (Schaefer (4), p. 237).

From 4.1.6 and the extension theorem 2.6.3, we have at once:

4.2.4. Let X be a locally convex, locally solid linear lattice, and let E be a linear sublattice of X. Then every continuous monotonic linear functional defined on E has a continuous monotonic extension to X.

Different topologies

In the spirit of 3.2.13, we have:

4.2.5. Suppose that X is a linear lattice, and that τ_1, τ_2 are topologies for X giving the same bounded sets. Suppose, also, that τ_1 is metrizable. Then:

(i) If the mapping $x \to x^+$ is continuous at 0 with respect to τ_2, then it is continuous at 0 with respect to τ_1.

(ii) If τ_2 is locally solid, then so is τ_1.

Proof. (i) By 4.1.2, if $x \to x^+$ is not τ_2-continuous at 0, then there is a bounded set A such that A^+ is unbounded. By 4.1.1, it follows that $x \to x^+$ is not τ_1-continuous at 0.

(ii) By 3.2.13, if τ_2 is locally order-convex, then so is τ_1. The result follows, by 4.1.5 (iv).

It is tempting to apply this result with τ_2 standing for the weak topology associated with τ_1, but the next (negative) result shows that, in many cases, the weak topology is not locally solid.

4.2.6. Let X be an infinite dimensional linear lattice, and let Y be an infinite dimensional subspace of X' that contains a strictly monotonic linear functional. Then the mapping $x \to x^+$ is not continuous at 0 with respect to $\sigma(Y)$.

Proof. Let f be a strictly monotonic linear functional in Y, and take elements $g_1, \ldots, g_n$ of Y. Then there exists a non-zero element x of X such that $f(x) = g_i(x) = 0$ $(i = 1, \ldots, n)$. Since f is strictly monotonic, $f(x^+) = f(x^-) > 0$. Taking a scalar multiple if necessary, we have $g_i(x) = 0$ $(i = 1, \ldots, n)$ while $f(x^+) > 1$.

Let τ be a locally convex, locally order-convex topology for an infinite-dimensional linear lattice X. By 3.4.3, $\sigma(X^*)$ is locally order-convex. If there is a strictly monotonic member of X*, the implication (iii) => (iv) of 4.1.5 now shows that P does not give an open decomposition of X with respect to $\sigma(X^*)$ (cf. example (iii) of section 3.3). This is the situation, for example, for m, c_o and l_1.

We now show that the two topologies considered in section 3.10 give an upper limit and (within a given dual pair) a lower limit for the locally convex, locally solid topologies on a linear lattice.

4.2.7. Let X be a linear lattice. Then the order-bound topology (τ_b) for X is locally solid, and is the largest locally convex, locally solid topology for X.

Proof. Any locally solid, locally convex topology makes order-intervals bounded, so is not larger than τ_b. Let U be a convex τ_b-neighbourhood of 0, and let V be the solid interior of U, that is $\{x: S(x) \subseteq U\}$. We show that V absorbs all order-intervals of the form $[0,a]$, from which it will follow, by 3.10.1, that V is a τ_b-neighbourhood of 0. Given $a > 0$, there exists $\delta > 0$ such that $\lambda[-a,a] \subseteq U$ for $0 \leq \lambda \leq \delta$. If $0 \leq x \leq a$ and $0 \leq \lambda \leq \delta$, then

$$S(\lambda x) = \lambda[x,x] \subseteq \lambda[-a,a] \subseteq U,$$

so $\lambda x \in V$, as required.

Let X be a linear lattice with positive cone P, and let Y be a solid subspace of X^b. Let P^o be evaluated in Y. By 2.6.5, for each f in P^o, we have

$$\sup \{g(x) : g \in Y \text{ and } -f \leq g \leq f\} = f(|x|) \quad (x \in X).$$

Hence the topology $\pi(Y)$ of uniform convergence on order-intervals of Y is given by the lattice seminorms p_f ($f \in P^o$), where $p_f(x) = f(|x|)$. This topology is therefore locally solid, and 3.10.9 shows that it is the smallest locally solid topology with dual space Y. If X^b contains an order-unit, then $\pi(X^b)$ coincides with τ_b (3.10.11), so this is the unique locally convex, locally solid topology for X having dual space X^b. Under the conditions of 4.2.6, of course, we know that $\pi(Y)$ is larger than $\sigma(Y)$.

4.3. M-spaces and pseudo-M-spaces

Let X be a topological linear space with a lattice ordering. We say that X satisfies condition (M) if there is a local base consisting of sublattices. If U is a sublattice, then so is $U \cap (-U)$, so the condition implies that there is a local base consisting of symmetric sublattices. It is sufficient if the neighbourhoods of 0 that admit $\vee$ form a local base, for then those that admit also do so; if we take a neighbourhood U that admits $\vee$, and V is a neighbourhood contained in U that admits $\wedge$, then $\vee(V)$ is a sublattice contained in U.

It is clear that condition (M) implies continuity at 0 of $x \to x^+$. However, it does not imply that this mapping is continuous at other points, as is shown by the lexicographic ordering of R^2 (since the ordering here is total, every subset is a sublattice).

Condition (M) makes sense in any topological space with a lattice ordering, but most of our results on the subject apply to locally convex spaces. A locally convex linear lattice will be called a <u>pseudo-M-space</u> if it satisfies condition (M), and an <u>M-space</u> if it is also locally solid. By 4.1.5, a pseudo-M-space is an M-space iff it is locally order-convex. In fact, local convexity then follows from the other assumptions, as the next result shows:

<u>4.3.1</u>. Let X be a linear lattice with a locally order-convex topology satisfying condition (M). Then there is a local base consisting of solid, convex sublattices.

<u>Proof</u>. Take order-convex U in Ⓝ(X). Then U contains a symmetric sublattice V in Ⓝ(X). By 2.3.12, [V] is a solid, convex sublattice.

It is easily seen that, for a linear lattice with a locally order-convex topology, condition (M) is equivalent to being locally directed.

<u>4.3.2</u>. In a pseudo-M-space, there is a local base consisting of convex sublattices.

<u>Proof</u>. This follows from the fact that the sublattice generated by a convex set is convex (2.2.10).

It is obvious that the Minkowski functional p of a convex sublattice U satisfies

$$p(x \vee y) \leq p(x) \vee p(y) \quad (x,y \in X) \qquad (1).$$

If U is also solid, then p is a lattice seminorm, so we have

$$p(x \vee y) = p(x) \vee p(y) \quad (x,y \in P) \qquad (2).$$

A lattice seminorm satisfying (2) will be called an <u>M-seminorm</u>. The relation $|x \vee y| \leq |x| \vee |y|$ shows that an M-seminorm satisfies (1). By 4.3.1, the topology of an M-space can be given by a family of M-seminorms. The same reasoning shows that the topology of a normable M-space can be given by one M-norm. By a slight abuse of language, we reserve the term "normed M-space" for a linear lattice with an M-norm (this is the traditional definition of an "abstract M-space", due to Kakutani (2)). If e is an order-unit, then $[-e,e]$ is a solid, convex sublattice, so the order-unit seminorm is an M-seminorm.

It is clear that, if a norm is given on a linear lattice, then condition (M) is satisfied iff the sublattice generated by the unit ball is bounded.

<u>Examples</u>

(i) The supremum norms on m and $C[0,1]$ are order-unit norms, and therefore M-norms. The supremum norm on c_0 is an M-norm, though not an order-unit norm.

(ii) The space s, with the usual topology and ordering, is an M-space. More generally, so is the space of all continuous real functions on a topological space S, when given the topology of uniform convergence on a specified family of compact sets. We shall see that every Hausdorff M-space can be embedded in a space of this type.

(iii) The space l_1 does not satisfy condition (M), since any sublattice containing the unit ball is unbounded.

(iv) Let m or c_0 have the usual norm and the ordering associated with the "partial-sum cone" P_s. For this ordering, $\|x \vee y\| \leq \|x\| \vee \|y\|$ (see 2.2), so the spaces are pseudo-M-spaces. They are not M-spaces, since P_s is not self-allied.

<u>Sublattices and compactness</u>

It is elementary that if condition (M) is satisfied, then the sublattice generated by a bounded set is bounded. In order to prove corresponding statements for totally bounded and compact sets, we start with the following lemma:

<u>4.3.3 LEMMA</u>. Suppose that X is an M-space and that $U \in Ⓝ(X)$. Then there exists V in $Ⓝ(X)$ such that if $n \in \omega$, $x_1, \ldots, x_n \in X$ and

$v_1, \dots, v_n \in V$, then

$$(x_1 + v_1) \vee \dots \vee (x_n + v_n) - (x_1 \vee \dots \vee x_n) \in U.$$

Proof. Let V be an order-convex sublattice in $\mathfrak{N}(X)$ that is contained in U. Write

$$v_1 \vee \dots \vee v_n = h, \quad v_1 \wedge \dots \wedge v_n = k,$$

$$x_1 \vee \dots \vee x_n = x, \quad (x_1 + v_1) \vee \dots \vee (x_n + v_n) = y.$$

Then $h, k \in V$ and $x + k \leq y \leq x + h$, so $y - x \in V$.

(Roughly speaking, this shows that, in an M-space, uniform continuity of the lattice operations extends simultaneously to all finite numbers of arguments.)

4.3.4. If X is an M-space, and A is a totally bounded subset of X, then $\vee(A)$ and the sublattice generated by A are totally bounded.

Proof. Given U in $\mathfrak{N}(X)$, take V in $\mathfrak{N}(X)$ with the property described in 4.3.3. There exist a finite number of points $a_1, \dots, a_n$ of A such that $A \subseteq \bigcup_{i=1}^{n} (a_i + V)$. Take x in $\vee(A)$. Then there exist points $x_1, \dots, x_k$ of A such that $x = x_1 \vee \dots \vee x_k$. For $1 \leq i \leq k$, there exists $r(i)$ such that $x_i - a_{r(i)} \in V$. Then

$$x - a_{r(1)} \vee \dots \vee a_{r(k)} \in U.$$

It follows that $\vee(A) \subseteq B + U$, where B denotes the (finite) set of suprema of non-empty subsets of $\{a_1, \dots, a_n\}$.

4.3.5. COROLLARY. If X is a complete, Hausdorff M-space, then every totally bounded subset of X has a supremum.

Proof. Let A be a totally bounded subset of X, and let B denote the closure of $\vee(A)$. The ordering is closed, so B has the same upper bounds as A. Now $\vee(A)$ is totally bounded, by 4.3.4, so B is compact. Since the mapping $(x,y) \to x \vee y$ is continuous, B admits $\vee$. Therefore, by 3.1.15, B has a greatest element.

These results apply equally to commutative groups.

Lattice homomorphisms

The basic theorem on the existence of continuous real lattice homomorphisms on pseudo-M-spaces has already been proved in a more general setting (3.1.12). For normed M-spaces, the following result (also derived from 1.8.2) will suit out pruposes better. It should be compared with 3.7.3.

4.3.6. Let X be a normed M-space, and let L_1 be the set of real lattice homomorphisms on X with the unit norm. Then, given x_o in X, there exists f in L_1 such that $|f(x_o)| = \|x_o\|$.

Proof. Let U be the closed unit ball in X, and let p be the Minkowski functional of U - P. Then $p(x) \leq \|x\|$ $(x \in X)$, so $p(x) \vee p(-x) \leq \|x\|$. If $\lambda > p(x) \vee p(-x)$, then x/λ and $-x/\lambda$ are in U - P, so there exist u,v in U such that $x/\lambda \leq u$, $-x/\lambda \leq v$. Then $|x|/\lambda \leq u \vee v \in U$, so $\|x\| \leq \lambda$. Hence $p(x) \vee p(-x) = \|x\|$. Given x_o in X, there exists, by 1.8.2, an element f of L_1 such that $f(x_o)=p(x_o)$. The result follows by applying this statement to x_o if $\|x_o\| = p(x_o)$ and to $-x_o$ if $\|x_o\| = p(-x_o)$.

Remarks (1) The space $L_1[0,1]$, with the usual norm and order, is a normed lattice, but has no non-zero real lattice homomorphisms at all. To prove this, we note first that, by 3.5.6, any monotonic linear functional must be continuous. Suppose that f is a real lattice homomorphism, and that $f(x) = 1$ for some x with $\|x\| = 1$. We can write x in the form $y_1 + y_1'$, where $\|y_1\| = \|y_1'\| = \frac{1}{2}$ and $y_1 \wedge y_1' = 0$. For one of y_1, y_1' (to be denoted by x_1), we must have $f(x_1) = 1$. Repeating the construction, we obtain a sequence of elements x_n such that $\|x_n\| = 2^{-n}$ and $f(x_n) = 1$, contradicting the continuity of f.

(2) As an application of 3.1.12, we can give an example of a real lattice homomorphism on an order-complete lattice which does not preserve countable infima (compare the example following 2.6.9, which was not order-complete). For the theorem tells us that there

is a real lattice homomorphism f on m such that f = 0 on c_0 and f(e) = 1. Let $a_n = e - (e_1 + \ldots + e_n)$. Then inf $\{a_n\} = 0$, but $f(a_n) = 1$ for each n.

Representation theorems

We now show how our results on the existence of real lattice homomorphisms can be used to represent M-spaces as spaces of continuous functions. As before, we denote by C(S) the space of all continuous functions on a topological space S.

(1) M-spaces

4.3.7 THEOREM (Jameson (1)). Let X be a Hausdorff M-space, and let L be the set of continuous real lattice homomorphisms on X, endowed with the topology $\sigma(X)$. For each x in X, a corresponding element $\hat{x}$ of C(L) is defined by : $\hat{x}(f) = f(x)$ $(f \in L)$. The mapping $x \to \hat{x}$ is a linear lattice isomorphism of X into C(L). It is a homeomorphism if C(L) is given the topology of uniform convergence on the sets $\{U^o \cap L : U \in \mathfrak{N}(X)\}$. If X is barrelled, this coincides with the topology of uniform convergence on compact sets.

Proof. The mapping $x \to \hat{x}$ is a linear lattice homomorphism, since this is true of each element of L. It is one-to-one, by 3.1.13, since the positive cone P is closed.

Now L is $\sigma(X)$-closed in X*, so $U^o \cap L$ is $\sigma(X)$-compact for each U in $\mathfrak{N}(X)$. Hence the topology (τ) for C(L) of uniform convergence on these sets is defined, and is not larger than the topology of uniform convergence on compact sets. To show that the mapping $x \to \hat{x}$ is a homeomorphism with respect to τ and the topology of X, we must prove that the sets $(U^o \cap L)^p$ $(U \in \mathfrak{N}(X))$ form a local base in X, where A^p denotes the absolute polar $\{x : |f(x)| \leq 1$ for f in A$\}$. Each such set is in $\mathfrak{N}(X)$, since $(U^o \cap L)^p \supseteq U \cap (-U)$. Conversely, suppose that U in $\mathfrak{N}(X)$ is given. By 3.2.4(i), there exists a symmetric sublattice V in $\mathfrak{N}(X)$ such that $\overline{V + P} \cap \overline{V - P} \subseteq U$. Take x not in U. If $x \notin \overline{V - P}$, then, by 3.1.12(i), there exists f in L such that $f \geq 1$ on V and

$f(x) > 1$. If $x \notin \overline{V + P}$, then, by 3.1.12(ii), there exists f in L such that $f \geq -1$ on V and $f(x) < -1$. In either case, $x \notin (V^o \cap L)^p$. Hence $(V^o \cap L)^p \subseteq U$.

If X is barrelled and K is a $\sigma(X)$-compact subset of L, then $K^p \in \mathfrak{N}(X)$. Since $K \subseteq K^{po} \cap L$, it follows that, in this case, τ coincides with the topology of uniform convergence on compact subsets of L.

If X is not barrelled, then τ may be strictly smaller than the topology of uniform convergence on compact sets. To show this, let F have the usual order and the supremum norm. For each n, the pointwise evaluation e_n is a lattice homomorphism. Further, $ne_n \to 0$ with respect to $\sigma(F)$, so E is $\sigma(F)$-compact, where $E = \{ne_n\} \cup \{0\}$. However, $E^p \notin \mathfrak{N}(X)$.

(2) Normed M-spaces

The next result follows immediately from 4.3.6:

4.3.8 (Kakutani (2)). Let X be a normed M-space, and let L_1 be the set of real lattice homomorphisms on X with unit norm. Then the natural mapping of X into $C(L_1)$ is an isometric linear lattice isomorphism.

For spaces with order-unit norms, this result can be sharpened:

4.3.9 THEOREM (Kakutani). Let X be a linear lattice with an order-unit norm, and let L_1 be the set of real lattice homomorphisms on X with unit norm, endowed with the topology $\sigma(X)$. Then L_1 is compact, and under the natural mapping, the image of X is dense in $C(L_1)$. If X is complete, then it can be identified with $C(L_1)$.

Proof. Let e be the order-unit. By 3.7.2, if $f \in P^o$ then $\|f\| = f(e)$. Hence L_1 is $\sigma(X)$-closed, and therefore $\sigma(X)$-compact. We use 0.3.11 to show that the image of X is dense in $C(L_1)$. We must show that, given distinct elements f,g of L_1, there exists x in X such that $f(x) \neq 0$ and $g(x) = 0$. Now $f(e) = g(e) = 1$, and there exists y in X such that $f(y) \neq g(y)$. Then $y - g(y)e$ is an element with the required property.

In a normed M-space without an order-unit, the set L_1 of real lattice homomorphisms with unit norm need not be $\sigma(X)$-compact. For instance, if X is c_o, then the pointwise evaluation e_n is in L_1 for each n, and $e_n \to 0$ with respect to $\sigma(X)$, so L_1 is not $\sigma(X)$-closed. Of course, the set L_o of real lattice homomorphisms f with $\|f\| \leq 1$ is $\sigma(X)$-compact, but the image of X under the natural mapping into $C(L_o)$ will always be a proper subset of $C(L_o)$, because of the compatibility conditions caused by the scalar multiplication present in L_o.

The Stone-Čech compactification

The theory of normed M-spaces gives a very natural approach to the Stone-Čech compactification of a topological space:

4.3.10 THEOREM. Let S be a completely regular, Hausdorff topological space. Then there exists a compact, Hausdorff space L such that

(i) S is homeomorphic to a dense subset of L, and

(ii) every bounded, continuous real-valued function on S has a continuous extension to L.

Proof. Let X be the space of bounded, continuous real-valued functions on S. Then X is complete with respect to the supremum norm, i.e. the order-unit norm associated ith the unit function. Let L be the set of real lattice homomorphisms on X with unit norm, endowed with the topology $\sigma(X)$. Then L is compact, and X is mapped onto C(L) by the natural mapping $x \to \hat{x}$, as in 4.3.9.

For each s in S, a corresponding element f_s of L is defined by: $f_s(x) = x(s) \quad (x \in X)$. If $s \neq t$, then there is an element x of X such that $x(s) \neq x(t)$, since S is Hausdorff and completely regular, so $f_s \neq f_t$. Now $\hat{x}(f_s) = x(s)$ $(x \in X)$, so, if we identify S with $\{f_s : s \in S\}$, then $\hat{x}$ defines an extension of x from S to L. It remains to show that the mapping $s \to f_s$ is a homeomorphism. Given a neighbourhood N of s in S, there exists x in X such that $x(s) = 1$ and $x(t) = 0$ for t not in N. Thus if $|f_t(x) - f_s(x)| < 1$, then $t \in N$. This shows that the mapping $f_s \to s$ is continuous. Conversely, a basic $\sigma(X)$-neighbourhood of f_s in L is of the form

$$U = \{g : |g(x_i) - f_s(x_i)| < \varepsilon \quad \text{for } i = 1,\dots,n\}.$$

There is a neighbourhood N of s such that $|x_i(s) - x_i(t)| < \varepsilon$ for t in N. Then $f_t \in U$ whenever $t \in N$, so the mapping $s \to f_s$ is continuous.

If the image of S is not dense in L, then there exists a non-zero element φ of C(L) that vanishes on $\{f_s : s \in S\}$. But each element of C(L) is $\hat{x}$ for some x in X, so this is impossible, and the image of S is dense in L.

If, in the above construction, we start with a compact, Hausdorff space S, then $\{f_s : s \in S\} = L$, so we have:

<u>4.3.11 COROLLARY</u>. (i) If S is a compact, Hausdorff space, then the only real lattice homomorphisms on C(S) are scalar multiples of the pointwise evaluation functionals.

(ii) If S_1, S_2 are compact, Hausdorff spaces, and there is an isometric linear lattice isomorphism between $C(S_1)$ and $C(S_2)$, then S_1 is homeomorphic to S_2.

<u>Proof</u>. (ii) Both S_1 and S_2 are homeomorphic to the space of unit real lattice homomorphisms on $C(S_1)$.

A direct proof of 4.3.11(i) goes as follows. Take $\varepsilon > 0$. Suppose that f is a real lattice homomorphism on C(S) and that, for each s in S, there exists x_s in C(S) such that $f(x_s) = 0$ and $x_s(s) > \varepsilon$. Then there is a neighbourhood N(s) of s such that $x_s(t) \geq \varepsilon$ for t in

$N(s)$. There exist $s_1,\ldots,s_n$ such that $\bigcup_{i=1}^{n} N(s_i) = S$. Let $x = x_{s_1} \vee \cdots \vee x_{s_n}$: then $x(t) \geq \varepsilon$ $(t \in S)$, so x is an order-unit in $C(S)$. But $f(x) = 0$, so $f = 0$. Hence there is an element s of S such that $x(s) = 0$ whenever $f(x) = 0$. It follows that the kernel of f is $\{x : x(s) = 0\}$, and that $f(x) = f(e)x(s)$ $(x \in X)$.

4.4. L-spaces

An L-space is a normed lattice in which the norm is additive on the positive cone. Typical examples are l_1, or, more generally, the L_1-space associated with any measure space (S,μ). It follows from 1.1.9(iv) that the conjugate of a normed M-space is an L-space. In an L-space, it is clear that there is a linear functional that agrees with the norm on the positive cone P, and that $\{x \in P : \|x\| = 1\}$ is a base for P. Hence the positive cone is well-based, and the results of section 3.8 apply; in particular, a complete L-space is order-complete. We now show that L-spaces are precisely the linear lattices with base norms.

4.4.1. (i) Let X be a linear lattice in which the positive cone has a base B, and let $\|\ \|$ be the corresponding base seminorm. Then $\|\ \|$ is a norm, and $(X, \|\ \|)$ is an L-space.

(ii) Every L-space has a base norm.

Proof. (i) By 3.9.1, $\|\ \|$ is additive on P. Therefore, for any x, $\|x\| \leq \|x^+\| + \|x^-\| = \|\,|x|\,\|$. We show that, conversely, $\|\,|x|\,\| \leq \|x\|$, from which it follows that $\|\ \|$ is a norm, since $\|x\| > 0$ for x in $P \sim \{0\}$. If $\|x\| < 1$, then there exist a,b in B and $\lambda,\mu > 0$ such that $\lambda+\mu < 1$ and $x = \lambda a - \mu b$. Then $|x| \leq \lambda a + \mu b$, so $\|\,|x|\,\| < 1$.

(ii) Let X be an L-space with positive cone P, and let

$B = \{ x \in P : \|x\| = 1 \}$. Then B is a base for P, and if p is the corresponding norm, then $p(x) = \|x\|$ $(x \in P)$. For any x in X, we have $\|x\| = \|x^+\| + \|x^-\|$, since X is an L-space, and $p(x) = p(x^+) + p(x^-)$, by (i). Hence $p(x) = \|x\|$ $(x \in X)$.

4.4.2 COROLLARY. The conjugate of an L-space is an M-space with an order-unit norm.

Proof. This follows from 3.9.4.

By the process of embedding in the second conjugate (see 2.6.10), this enables us to embed every normed M-space in one with an order-unit norm, and every L-space in the dual of a space C(S), where S is compact and extremally disconnected. Now the dual of C(S) can be identified with the space of signed, regular Borel measures on S, with variation as norm. The L-space can be identified with the set of measures that vanish on Borel sets of the first category (see Kelley-Namioka (1), section 24). Alternatively, a direct (but lengthy) construction can be performed to show that every L-space is isomorphic to a space $L_1(S,\mu)$ (see Kakutani (1)). Instead of reproducing any of these proofs here, we mention some more properties which follow quite easily from the definition.

4.4.3. If a non-trivial linear lattice is both an L-space and a normed M-space, then it is one-dimensional.

Proof. We show that the space is totally ordered, from which the result follows, by 1.5.11. Take x,y in X. Let $z = x^- + y^-$, and write $a = x + z$, $b = y + z$. Then $a,b \geq 0$. Suppose that $\|a\| \leq \|b\|$. Then $\|a \vee b\| = \|b\|$, since X is a normed M-space, and $\|a \vee b\| = \|(a \vee b) - b\| + \|b\|$, since X is an L-space. Hence $a \vee b = b$, so $a \leq b$, and $x \leq y$.

By 3.8.11 and 4.4.1, if X is a linear lattice in which the positive cone is well-based and gives an open decomposition, then there is a norm inducing the topology of X and making it an L-space.

We recall from section 3.8 that any normed space contains plenty of well-based cones with interior points. Our next result, however, shows that it is not easy for such cones to give an ordering satisfying the Riesz interpolation property.

4.4.4. If the positive cone in a Hausdorff Riesz space X is well-based and has an interior point, then X is finite dimensional.

Proof. We show that X* is finite dimensional. The topology of X can be given by an order-unit norm, so that of X* can be given by the base norm corresponding to a $\sigma(X)$-compact base B. Also, the positive cone P^0 in X* has an interior point f_0, by 3.8.4. Let E be the set of extreme points of B. By the Krein-Mil'man theorem, B is the $\sigma(X)$-closure of co(E). We show that E is finite, from which it follows that B, and therefore X*, is contained in the linear span of E. Suppose, then, that E is infinite, and take a sequence $\{f_n\}$ of distinct members of E. By 2.2.14, $\{f_n\}$ is pairwise disjoint, so if $g_n = f_1 \vee \dots \vee f_n$, then $g_n = f_1 + \dots + f_n$, and $\|g_n\| = \|f_1\| + \dots + \|f_n\| = n$. But there is a positive number α such that $f \leq \alpha f_0$ $(f \in B)$, so $g_n \leq \alpha f_0$ and $\|g_n\| \leq \alpha\|f_0\|$ for each n. This is a contradiction.

CHAPTER 5

ORDERED ALGEBRAS

As pointed out in the introduction, this short final chapter is designed to give a taste, rather than a treatment, of ordered algebras. The literature on this subject is somewhat scattered, containing an interesting, but diffuse, collection of methods and results. Many of the results reproduced here have been generalised, and others probably could be. In some ways, the effect of a multiplication is similar to the effect of a lattice ordering. In particular, the multiplicative linear functionals can be characterised as extreme points of various sets, and this leads to a representation theorem comparable to those applying to M-spaces. Matters related to this occupy the first two sections; the third section is devoted to finding conditions under which the spectral radius of an element of a Banach algebra is in its spectrum, thereby generalising some classical results on eigenvalues.

5.1. Basic results

Let X be an associative linear algebra over the real field. A wedge in X that admits multiplication will be called a <u>semi-algebra</u>. A semi-algebra will be said to be <u>strict</u> if it is a cone. Clearly, the ordering associated with a semi-algebra is preserved by multiplication by positive elements in the following sense: if $x \leqslant y$ and $z \geqslant 0$, then $xz \leqslant yz$ and $zx \leqslant zy$. The least semi-algebra containing an element x is the set of polynomials in x with non-negative coefficients and no constant term. If the algebra has an identity e, and constant terms are admitted to the polynomials, we obtain the least semi-algebra containing x and e.

Three natural examples of orderings in algebras come to mind:

(i) In an algebra of real-valued functions, the usual positive cone is a semi-algebra.

(ii) If X is an ordered linear space, and Y is an algebra of linear mappings of X into itself, then the set of monotonic linear mappings in Y is a semi-algebra.

(iii) In any algebra, the set P of finite sums of squares is a wedge. In a commutative algebra, P is a semi-algebra. Clearly, P coincides with the usual positive cone in the algebra of all real-valued functions on a set S. If the algebra has an identity, then P generates it, since, for any x,

$$4x = (x + e)^2 - (x - e)^2.$$

This, however, is not true in an algebra without an identity: if l_2 is given the pointwise multiplication, then all squares lie in the proper subspace l_1.

Some (but not all) of our results apply to Banach algebras. The next theorem summarises the basic properties that will be needed. Proofs can be found in Rickart (1).

<u>5.1.1</u>. Let X be a Banach algebra with an identity e. Then:

(i) For each x in X, $\lim_{n\to\infty} \|x^n\|^{1/n}$ exists. (This number is called the <u>spectral radius</u> of x, and will be denoted by $\rho(x)$).

(ii) If $\rho(x) < 1$, then $(e - x)^{-1}$ exists, and is equal to $e + \sum_{n=1}^{\infty} x^n$.

(iii) If $\rho(x) < 1$, then there exists y such that $y^2 = e - x$.

Part (iii) of 5.1.1, shows that, in a Banach algebra with an identity, the identity is an interior point of the wedge of sums of squares. We now mention some properties of linear functionals that are non-negative on this wedge.

<u>5.1.2</u>. Let X be an algebra, and let P be a wedge in X containing all squares. Let f be an element of P^o. If a,b are commuting elements of X, then

$$(f(ab))^2 \leq f(a^2)\, f(b^2).$$

If X has an identity e, then

$$(f(a))^2 \leq f(a^2)\, f(e) \quad (a \in X),$$

and $\{g \in P^o : g(e) = 1\}$ is a base for P^o.

<u>Proof</u>. For any real number λ, we have

$$0 \leq f((\lambda a + b)^2)$$
$$= \lambda^2 f(a^2) + 2\lambda f(ab) + f(b^2).$$

The stated inequalities follow. If $f(e) = 0$, the second inequality shows that $f = 0$.

<u>5.1.3</u>. Let X be a Banach algebra with an identity e, and let f be a linear functional on X that is non-negative on squares. Then $|f(x)| \leq \rho(x)\, f(e)$ $(x \in X)$. In particular, f is continuous.

<u>Proof</u>. If $\rho(x) < 1$, then, by 5.1.1 (iii), there exists y in X such that $y^2 = e - x$. Hence $f(x) = f(e - y^2) \leq f(e)$. The result follows, since $\rho(\lambda x) = |\lambda|\rho(x)$ for all real λ.

We say that a linear mapping f from one algebra to another is <u>multiplicative</u> if $f(xy) = f(x)\, f(y)$ holds for all x,y. It is clear that a multiplicative linear functional is positive on squares (so satisfies the conclusions of 5.1.2 and 5.1.3). It is worth noting, however, that a multiplicative linear functional on an algebra of real functions need not be monotonic with respect to the natural ordering. For instance, let the algebra of polynomials be ordered as functions on $[0,1]$, and define $f(x) = x(2)$.

<u>5.1.4 LEMMA</u>. Let X,Y be algebras, and let P be a semi-algebra that generates X. If f is a linear mapping from X to Y, and $f(ab) = f(a)f(b)$ for a,b in P, then f is multiplicative.

<u>Proof</u>. Straightforward.

Stone's theorem

5.1.5 THEOREM (Stone). Let X be an algebra with an identity e and an ordering given by a semi-algebra P. Suppose that e is also an order-unit, and let p denote the corresponding order-unit seminorm. Then:

(i) If f is an extremal monotonic linear functional and f(e) = 1, then f is multiplicative.

(ii) There is a compact, Hausdorff space E and a monotonic, multiplicative linear mapping $x \to \hat{x}$ of X into C(E) such that

$$\sup\{|\hat{x}(s)| : s \in E\} = p(x) \quad (x \in X)$$

and $\{\hat{x} : x \in X\}$ is dense in C(E) with respect to the supremum norm.

(iii) If the ordering is almost Archimedean, then X is commutative.

Proof. (i) By 5.1.4, it is sufficient to show that f(ax) = f(a)f(x) for a in P, x in X. Take a in P, and let g(x) = f(ax) (x ∈ X). Then g(x) ≥ 0 (x ∈ P). Now a ≤ λe for some λ > 0, so ax ≤ λx (x ∈ P), and g(x) ≤ λf(x) (x ∈ P). Hence 0 ≤ g ≤ λf, so there exists μ > 0 such that g = μf. But g(e) = f(a), and f(e) = 1, so μ = f(a).

(ii) Let E be the set of multiplicative, monotonic linear functionals that take the value 1 at e (that is, all non-zero ones). Then E is σ(X)-compact. By 3.7.3, given x in X, there exists f in E such that f(x) = p(x). For each x in X, define a corresponding element $\hat{x}$ of C(E) by: $\hat{x}(f) = f(x)$ (f ∈ E). Then the mapping $x \to \hat{x}$ is clearly monotonic, multiplicative and linear, and it folows from 0.3.11 and 0.3.12 that its range is dense in C(E).

(iii) If the ordering is almost Archimedean, then p is a norm, so the mapping $x \to \hat{x}$ is one-to-one. Since C(E) is commutative, it follows that X is.

A generalisation of the above theorem to spaces with approximate order-units and identities is given in Ng (1).

Stone's theorem indicates that there is no very great loss of generality in considering algebras of real functions. There is quite an extensive literature on semi-algebras of real functions, especially ones satisfying a condition of the following form: if $x \in P$, then $x^n/(1 + x) \in P$. For an introduction to this theory, see Bonsall (4), (6), Pryce (1).

5.2. Multiplicative and extreme monotonic linear mappings

Motivated by Stone's theorem, it is reasonable to look for further results identifying multiplicative linear mappings with extreme points of various sets of monotonic ones. Firstly, by a more careful application of the method used in 5.1.5, we obtain:

5.2.1. Let X be a commutative algebra ordered by a semi-algebra P containing order-units, and let Y be an algebra ordered by a wedge Q. If f is an extremal monotonic linear mapping from X to Y, then

either (i) $f(xy) = 0$ for all x,y in X,

or (ii) there exist a monotonic, multiplicative linear functional h on X and an extremal element y_o of Q such that $f(x)=h(x)y_o$ $(x \in X)$.

Proof. Take a in P. Reasoning as in 5.1.5(i), we see that there is a non-negative number h(a) such that $f(ax) = h(a)\, f(x)$ $(x \in X)$. Since X has order-units, P generates X, and for each a in X there is a number h(a) such that this relation holds. The functional h so defined is clearly linear. Choose x such that $f(x) \neq 0$. For any a,b in X, we have

$$\begin{aligned} h(ab)\, f(x) &= f(abx) \\ &= h(a)\, f(bx) \\ &= h(a)\, h(b)\, f(x). \end{aligned}$$

Hence h is multiplicative. If $h = 0$, case (i) occurs. Otherwise, there exists a in P such that $h(a) = 1$. Then

$$f(x) = f(ax) = f(xa) = h(x)\, f(a) \quad (x \in X).$$

From the fact that f is extremal, it follows in the usual way that f(a) is an extremal point of Q.

Clearly, alternative (i) cannot occur for a non-zero mapping if X has an identity. To give an example where it can occur, consider the algebra of polynomials p of the form $p(t) = t^2q(t)$. A non-zero linear functional f on this algebra that satisfies (i) is defined by: $f(p)=q(0)$.

For linear functionals, we have a converse result:

5.2.2. Let X be an algebra with an identity e, and let P be a wedge in X containing all squares. Then each multiplicative linear functional in P^o is an extremal element of P^o.

Proof. Let $B = \{f \in P^o : f(e) = 1\}$. We have seen (5.1.2) that B is a base for P^o, so it is sufficient to show that if f is a non-zero, multiplicative functional in P^o, then f is an extreme point of B. Certainly, $f \in B$. Suppose that g,h are elements of B such that $f = \frac{1}{2}(g + h)$. Then, for any x in X,

$$f(x^2) = \tfrac{1}{2}g(x^2) + \tfrac{1}{2}h(x^2),$$

$$(f(x))^2 = \tfrac{1}{4}(g(x))^2 + \tfrac{1}{2}g(x)h(x) + \tfrac{1}{4}(h(x))^2.$$

By 5.1.2, $g(x^2) \geq (g(x))^2$ and $h(x^2) \geq (h(x))^2$, so

$$(g(x) - h(x))^2 = (g(x))^2 - 2g(x)h(x) + (h(x))^2 \leq 0.$$

Hence $g(x) = h(x)$, so $g = h = f$.

Combining 5.2.1 and 5.2.2, we see that if X is a commutative algebra with an identity, and if P is a semi-algebra in X that contains all squares and has internal points, then the non-zero multiplicative elements of P^o are precisely the extreme points of B.

When we consider linear mappings (rather than functionals), it is evident that there will usually be plenty of multiplicative, monotonic ones that are not extremal monotonic - for instance, the identity mapping in a non-trivial algebra. The rest of this section is devoted to showing that, for certain algebras of real-valued functions (with the usual operations and order), the multiplicative, monotonic linear mappings can be identified with the extreme points of a suitable set of monotonic linear mappings. Even when not assuming that the algebra in question contains the constant functions, we shall write $x \leq 1$ to mean $x(s) \leq 1$ for all s. Since $0 \leq x \leq 1$ is equivalent to $0 \leq x^2 \leq x$, a multiplicative, monotonic linear mapping f has the property that $0 \leq f(x) \leq 1$ whenever $0 \leq x \leq 1$: this suggests what might be a "suitable" set of monotonic linear mappings. We say that an algebra X of real functions is <u>positively generated</u> if the cone of non-negative functions in X is generating.

<u>5.2.3</u>. Let X be an algebra of bounded real functions, and let Y be any algebra of real functions. Let K_o be the set of monotonic linear mappings f from X to Y for which $0 \leq f(x) \leq 1$ whenever $0 \leq x \leq 1$. Then every extreme element of K_o is multiplicative.

<u>Proof</u>. Let f be an extreme element of K_o. Take y in X with $0 \leq y \leq 1$, and define

$$g(x) = f(xy) - f(x)\,f(y) \qquad (x \in X).$$

We show that f + g and f - g are both in K_o, so that g = 0. The result then follows, since functions in X are bounded. If $x \geq 0$, then

$$f(x) + g(x) = f(x)(1 - f(y)) + f(xy) \geq 0,$$

since $f(y) \leq 1$, and

$$f(x) - g(x) = f(x - xy) + f(x)\,f(y) \geq 0,$$

since $x \geq xy$. If $0 \leq x \leq 1$, then $0 \leq f(x) \leq 1$ and $0 \leq xy \leq y$, so

$$f(x) + g(x) = f(x)(1 - f(y)) + f(xy)$$

$$\leq (1 - f(y)) + f(y) = 1,$$

and $\qquad f(x) - g(x) \leq f(x) - f(xy) + f(y) \leq 1,$

since $\qquad x - xy + y \leq 1 - y + y = 1.$

Under the hypothesis that X contains the constant functions, we can use 5.2.2 to prove the converse:

<u>5.2.4</u>. Let X be an algebra of real functions that is positively generated and contains the constant functions, and let Y be any algebra

of real functions. Let K_0 be the set of monotonic linear mappings f from X to Y for which $f(e) \leq 1$. Then every multiplicative, monotonic linear mapping from X to Y is an extreme point of K_0.

<u>Proof</u>. Since X is positively generated, 0 is an extreme point of K_0. Let f be a non-zero multiplicative, monotonic linear mapping. By the remarks preceding 5.2.3, $f \in K_0$. Suppose that f + g and f - g are in K_0. Denote by T the domain of functions in Y. For each t in T, let $\varphi_t(y) = y(t)$ $(y \in Y)$. Then $\varphi_t \circ f$ is a multiplicative, monotonic linear functional on X, so $(\varphi_t \circ f)(e)$ is 1 or 0. Also, $\varphi_t \circ (f \pm g)$ are monotonic linear functionals on X, with values not greater than 1 at e. If $(\varphi_t \circ f)(e) = 0$, then $\varphi_t \circ f = 0$, by 5.1.2, so $\varphi_t \circ g = 0$. If $(\varphi_t \circ f)(e) = 1$, then 5.2.2 shows that $\varphi_t \circ f$ is an extreme point of the set of monotonic linear functionals taking the value 1 at e, so, again, $\varphi_t \circ g = 0$. Hence $\varphi_t \circ g = 0$ for all t, so $g = 0$.

For further results of this kind, see Bonsall-Lindenstrauss-Phelps (1). Among other things, it is shown there that 5.2.4 holds under the hypothesis that X is an algebra of bounded functions (not necessarily containing the constant functions), and that both 5.2.3 and 5.2.4 hold under the alternative hypothesis that if x is a positive element of X, then $x/(1 + x) \in X$. It is also proved that the linear functional mentioned after 5.2.1 is, in fact, extremal monotonic. Further developments are to be found in Converse (1).

<u>5.3. Spectral properties of positive elements</u>

If x is an element of a real algebra X with identity e, then a real number λ is said to be in the <u>spectrum</u> of x if $\lambda e - x$ has no inverse. Our concern in this section is to find conditions on an element x of a Banach algebra under which the spectral radius $\rho(x)$ is in the spectrum of x. This is necessarily true if $\rho(x) = 0$. We shall prove that if $\rho(x) > 0$, then $\rho(x)$ is in the spectrum of x if (1) there is a closed, self-allied semi-algebra containing x and e, or (2) x is an element giving compact multiplication and there is a strict closed semi-algebra containing x and e. In case (2), we will in fact prove the stronger result that $\rho(x)e - x$ is a zero-divisor. These results were obtained independently by Bonsall and Schaefer; the methods reproduced here are those of Bonsall. We shall require the following elementary facts concerning inverses in a Banach algebra (we outline the proofs):

5.3.1. Let X be a Banach algebra with identit e, and let x be an element of X. Let u_λ denote the inverse of $\lambda e - x$ whenever this exists. Then:

(i) If $\lambda > \rho(x)$, then u_λ exists and is equal to

$$\lambda^{-1}e + \lambda^{-2}x + \lambda^{-3}x^2 + \dots .$$

(ii) If u_λ exists and $|\mu-\lambda| < \|u_\lambda\|^{-1}$, then u_μ exists, and

$$u_\mu = u_\lambda + (\lambda-\mu)u_\lambda^2 + (\lambda-\mu)^2 u_\lambda^3 + \dots ,$$

$$\|u_\mu - u_\lambda\| < (1 - |\mu-\lambda| \, \|u_\lambda\|)^{-1} |\mu-\lambda| . \|u_\lambda\|^2 .$$

(iii) $\rho(x)$ is in the spectrum of x iff $\{u_\lambda : \lambda > \rho(x)\}$ is unbounded.

(iv) If u_λ exists, then

$$u_\lambda = \lambda^{-1}e + \lambda^{-2}x + \dots + \lambda^{-n}x^{n-1} + \lambda^{-n}x^n u_\lambda .$$

Proof. (i) and (ii) follow easily from 5.1.1(ii). (For (ii), write $\mu e - x = (\mu-\lambda)e + (\lambda e - x) = (\lambda e - x)(e - (\lambda-\mu)u_\lambda)$.)

(iii) Suppose that $\rho(x)e - x$ is invertible. Take $\lambda_0 > \|x\|$. It follows from (i) that $\{u_\lambda : \lambda > \lambda_0\}$ is bounded. By (ii), the mapping $\lambda \to \|u_\lambda\|$ is continuous, so $\{u_\lambda : \rho(x) \leq \lambda \leq \lambda_0\}$ is also bounded.

Conversely, suppose that $\|u_\lambda\| \leq \alpha$ for $\lambda > \rho(x)$. If $\mu = \rho(x) + \frac{1}{2\alpha}$ then $\|u_\mu\| \leq \alpha$, so, by (ii), $u_{\rho(x)}$ exists.

(iv) This follows on multiplying by $\lambda e - x$.

The following lemma is basic to the theory:

5.3.2. LEMMA. Let X be a Banach algebra with identity e, and let P be a closed semi-algebra containing e. Let x be an element of P such that $\rho(x) > 0$, and let u_λ denote the inverse of $\lambda e - x$ whenever this exists. If $u_{\rho(x)}$ exists, then there exists μ such that $0 < \mu < \rho(x)$, u_μ exists, and, for each n in ω,

$$u_\mu = \mu^{-n-1}x^n + q_n,$$

where $q_n \in P$.

Proof. By 5.3.1 (iii), there exists $\alpha > 0$ such that $\|u_\lambda\| \leq \alpha$ for all $\lambda > \rho(x)$. By 5.3.1 (i), $u_\lambda \in P$ for $\lambda > \rho(x)$. Take λ, μ such that $\lambda > \rho(x)$, $0 < \mu < \rho(x)$ and $\lambda - \mu < \alpha^{-1}$. By 5.3.1 (ii), u_μ exists and is equal to

$$u_\lambda + (\lambda-\mu)u_\lambda^{\,2} + (\lambda-\mu)u_\lambda^{\,3} + \ldots ,$$

so $u_\mu \in P$. By 5.3.1 (iv),

$$\begin{aligned} u_\mu &= \mu^{-1}e + \mu^{-2}x + \ldots + \mu^{-n-1}x^n + \mu^{-n-1}x^{n+1}u_\mu \\ &= \mu^{-n-1}x^n + q_n, \end{aligned}$$

where $q_n \in P$, since $u_\mu \in P$.

Elements of self-allied semi-algebras

5.3.3 THEOREM. Let X be a Banach algebra with identity e, and let P be a closed, self-allied semi-algebra containing e. Let x be an element of P such that $\rho(x) > 0$. Denote the inverse of $\lambda e - x$ by u_λ whenever it exists. Then:

(i) $\rho(x)$ is in the spectrum of x.

(ii) $u_\lambda \in P$ iff $\lambda > \rho(x)$.

Proof. (i) If $\rho(x)$ is not in the spectrum of x, then there exist μ and q_n as in 5.3.2. Since P is self-allied, $\{\mu^{-n-1}x^n\}$ is bounded. But this implies that $\mu \geq \rho(x)$, contrary to hypothesis.

(ii) We notice first that if x is invertible, then its inverse is not in -P, for otherwise -e would be in -P, which is not so, since $e \in P$ and P is strict.

If $\lambda > \rho(x)$, then $u_\lambda \in P$, by 5.3.1(i). Suppose now that $\lambda \neq 0$ and $u_\lambda \in P$. Then

$$\lambda u_\lambda = e + xu_\lambda \in P,$$

so $\lambda > 0$. Suppose that $\lambda < \rho(x)$. Take $\mu > \rho(x)$. Then, with respect to the ordering associated with P,

$$\begin{aligned} u_\lambda &= \lambda^{-1}e + \dots + \lambda^{-n}x^{n-1} + \dots \\ &\geq \mu^{-1}e + \dots + \mu^{-n}x^{n-1} + \dots \\ &= u_\mu, \end{aligned}$$

so $0 \leq u_\mu \leq u_\lambda$. But this implies that $\{u_\mu : \mu > \rho(x)\}$ is bounded, and therefore that $\rho(x)$ is not in the spectrum of x.

5.3.4. COROLLARY. Let X be a Banach space, ordered by a closed, generating, self-allied cone P. If f is a monotonic linear mapping of X into itself, and $\rho(f) > 0$, then $\rho(f)$ is in the spectrum of f.

Proof. Let Q be the set of monotonic linear mappings of X into itself. Then Q is a semi-algebra containing the identity. By 3.5.2, P gives an open decomposition of X, and by 3.5.5, each element of Q is continuous. By 3.6.3, Q is self-allied. The result follows.

Elements giving compact multiplication

5.3.5 THEOREM. Let X be a Banach algebra with identity e, and let x be an element of X such that (i) $\rho(x) = \rho > 0$, (ii) the mapping $y \to xy$ $(y \in X)$ is compact, and (iii) the least closed semi-algebra P containing x and e is strict. Then there is a non-zero element v of P such that $xv = \rho v$.

Proof. Suppose that $\rho e - x$ is invertible. Then, by 5.3.2, there exists μ such that $0 < \mu < \rho$, u_μ exists, and

$$u_\mu = \mu^{-n-1}x^n + q_n,$$

where $q_n \in P$ for each n. Let $\alpha_n = \mu^{-n-1}\|x^n\|$. Then $\alpha_n \to \infty$ since $\mu < \rho$ so there is a subsequence $\{\alpha_{n_k}\}$ such that $\alpha_{n_k} > \alpha_{n_k-1}$ for each k. Write

$$a_k = \mu^{-n_k}x^{n_k-1}.$$

Then $\|a_k\| = \alpha_{n_k-1} \to \infty$, and

$$u_\mu = \mu^{-1}xa_k + q_{n_k} \quad (k = 1,2, \ldots).$$

Hence

$$\|a_k\|^{-1}u_\mu = \mu^{-1}x \ \|a_k\|^{-1}a_k + \|a_k\|^{-1} q_{n_k} \qquad (1).$$

Since the mapping $y \to xy$ is compact, there is a subsequence of $\{\mu^{-1}x\|a_k\|^{-1} \ a_k\}$ that has a limit, say b. Clearly, $b \in P$. From (1), we see that the corresponding subsequence of $\{\|a_k\|^{-1} q_{n_k}\}$ converges to $-b$, showing that $-b \in P$. Since P is strict, it follows that $b = 0$. But $\|\mu^{-1}xa_k\| = \alpha_{n_k} > \|a_k\|$ for each k, so $\|b\| \geq 1$. This is a contradiction, so $\rho e - x$ is not invertible, and there is a sequence $\{\beta_n\}$ such that $\beta_n > \rho$ for each n, $\beta_n \to \rho$ and $\|u_{\beta_n}\| \to \infty$. Let $v_n = \|u_{\beta_n}\|^{-1} u_{\beta_n}$.

Then $v_n \in P$ for each n, and there is a subsequence $\{v_{n_k}\}$ such that $\{xv_{n_k}\}$ converges to an element v of P.

Now

$$(\beta_n e - x)v_n = \|u_{\beta_n}\|^{-1} e \to 0,$$

so $\beta_{n_k} v_{n_k} \to v$. Hence $v \neq 0$ and

$$xv = \lim_{k\to\infty} \ (\beta_{n_k} \ x \ v_{n_k}) = \rho v.$$

It can be shown (cf. Bonsall-Tomiuk (1)) that the semi-algebra P in 5.3.5 is locally compact, and therefore self-allied. However, the proof of this requires the result of 5.3.5 first, so does not help us to deduce 5.3.5 from 5.3.3. For an introduction to the theory of locally compact semi-algebras, see Bonsall (5).

Again we show that there is a corresponding result applying to linear mappings on a Banach space. The next lemma is the key to it. We denote by B(X) the algebra of bounded linear mappings of X into itself.

<u>5.3.6 LEMMA</u> (Bonsall). Let X be a Banach space, and let f be a compact linear mapping of X into itself. Let C be the centraliser of f in B(X). Then the mapping $g \to fg$ $(g \in C)$ is compact.

<u>Proof</u>. Let X_1, C_1 denote the closed unit balls in X,C, and let

$E = f(X_1)$. Then E is totally bounded, and for g in C_1, we have

$$g(E) = gf(X_1) = fg(X_1) \subseteq f(X_1) = E.$$

By Ascoli's theorem (0.3.10), it follows that, for each $\varepsilon > 0$, there is a finite subset $\{g_1, \ldots, g_n\}$ of C_1 such that, given g in C_1, there exists i such that

$$\|g(x) - g_i(x)\| < \varepsilon \qquad (x \in E),$$

so that $\|gf - g_i f\| < \varepsilon$. Hence $\{gf : g \in C_1\}$ is totally bounded, which proves the result.

(It is not true that if f is a compact linear mapping, then the mapping $g \to fg$ is compact on the whole of $B(X)$, but this is true of the mapping $g \to fgf$; see Bonsall (7), Alexander (1)).

<u>5.3.7 THEOREM</u> (Krein-Rutman). Let X be a Banach space, and let P be a closed cone such that $\overline{P - P} = X$. Let f be a compact linear mapping of X into itself such that $f(P) \subseteq P$ and $\rho(f) = \rho > 0$. Then there exist non-zero elements x_0 of P and h_o of P^o such that $f(x_0) = \rho x_0$ and $f^*(h_o) = \rho h_o$ (where f^* denotes the adjoint of f).

<u>Proof</u>. Let C be the centraliser of f in $B(X)$, and let $Q = \{g \in C : g(P) \subseteq P\}$. Then Q is a strict, closed semi-algebra containing f and the identity, and the mapping $g \to fg$ $(g \in C)$ is compact. Therefore, applying 5.3.5 to C, there is a non-zero element g of Q such that $fg = \rho g$. Since g is non-zero, there exists x_1 in P such that $g(x_1) \neq 0$: let $x_0 = g(x_1)$. Then $x_0 \in P \sim \{0\}$, and

$$f(x_0) = fg(x_1) = \rho g(x_1) = \rho x_0.$$

Since $x_0 \notin -P$, there exists h_1 in P^o such that $h_1(x_0) = 1$. Let $h_o = g^*(h_1)$. Then $h_o \in P^o$, and $h_o \neq 0$, since

$$h_o(x_1) = h_1(g(x_1)) = h_1(x_0) = 1.$$

Finally, $f^*g^* = g^*$ (since $gf = fg = \rho g$), so

$$f^*(h_o) = \rho g^*(h_1) = \rho h_o.$$

APPENDIX. CS-closed sets

Let A be a subset of a Hausdorff topological linear space. By a <u>convex series</u> of elements of A we mean a series of the form $\sum_{n=1}^{\infty} \lambda_n a_n$, where $a_n \in A$ and $\lambda_n \geq 0$ for each n, and $\sum_{n=1}^{\infty} \lambda_n = 1$. We shall use the following terminology:

(i) A is <u>CS-closed</u> if it contains the sum of every convergent convex series of its elements;

(ii) A is <u>CS-compact</u> if every convex series of its elements converges to a point of A.

Clearly, every CS-compact set is CS-closed, and every CS-closed set is convex.

It is elementary that every sequentially closed, convex set is CS-closed, and that every sequentially complete, bounded convex set is CS-compact. Conversely, every CS-compact set is bounded. Many non-closed convex sets are CS-closed, for example the open unit ball in a normed space. A wedge is CS-closed iff it is series-closed. An example of a non-closed wedge with this property is the positive cone for the lexicographic ordering in any of the usual sequence spaces.

Let us say that a set A is <u>semi-closed</u> if A and its closure $\overline{A}$ have the same interior.

A.1. THEOREM. Every CS-closed subset of a metrizable topological linear space is semi-closed.

<u>Proof</u>. Suppose that A is CS-closed. It is sufficient to prove that $\text{int}\ \overline{A} \subseteq A$. Let x be an interior point of $\overline{A}$, and write $B = A - x$. Then B is CS-closed, and $\overline{B}$ is a neighbourhood of 0. We must show that $0 \in B$.

Let $\{U_n : n = 1, 2, \ldots\}$ be a decreasing base of neighbourhoods of 0. For each n, let

$$V_n = U_n \cap (2^{-n}\,\overline{B}).$$

Then each V_n is a neighbourhood of 0. We now construct a sequence

$\{b_n\}$ in B as follows. Since $0 \in \frac{1}{2}\overline{B}$ and V_2 is a neighbourhood of 0, there is a point b_1 of B such that $-\frac{1}{2}b_1 \in V_2$. Write $y_o = 0$, $y_1 = \frac{1}{2}b_1$. Suppose now that $b_1,\ldots,b_{n-1}$ and $y_o, y_1,\ldots,y_{n-1}$ have been defined so that $b_i \in B$, $-y_i \in V_{i+1}$ and $y_i = y_{i-1} + 2^{-i}b_i$ for each i. Then $-y_{n-1} \in 2^{-n}\overline{B}$, so there is a point b_n of B such that $-y_n \in V_{n+1}$, where $y_n = y_{n-1} + 2^{-n}b_n$. Now $y_n \to 0$, and

$$y_n = 2^{-1}b_1 + \ldots + 2^{-n}b_n,$$

so $\sum_{n=1}^{\infty} 2^{-n}b_n = 0$. Since B is CS-closed, it follows that $0 \in B$, as required.

The generalisation of Tukey's theorem required in 3.6 now follows from:

__A.2.__ Suppose that A is CS-compact and B is CS-closed. Then A + B and co(A ∪ B) are CS-closed.

__Proof__. We prove the result for co(A ∪ B). The result for A + B can be proved similarly. Let $\sum_{n=1}^{\infty} \lambda_n x_n$ be a convergent convex series of elements of co(A ∪ B), with sum x. For each n, there exist a_n in A, b_n in B and μ_n in $[0,1]$ such that $x_n = \mu_n a_n + (1 - \mu_n)b_n$. Let $\mu = \sum_{n=1}^{\infty} \lambda_n \mu_n$. Then $\mu \in [0,1]$, and $\sum_{n=1}^{\infty} \lambda_n(1 - \mu_n) = 1 - \mu$. Since A is CS-compact, $\sum_{n=1}^{\infty} \lambda_n \mu_n a_n = \mu a$, where $a \in A$. Hence $\sum_{n=1}^{\infty} \lambda_n(1 - \mu_n)b_n$ is convergent, and since B is CS-closed, its sum is $(1 - \mu)b$ for some b in B. Thus $x = \mu a + (1 - \mu)b \in$ co (A ∪ B).

Even if X is not metrizable, the following is true: if A is CS-compact and B is closed and convex, then A + B and co(A ∪ B) are semi-closed. For a proof of this and other results, see Jameson (5).

It is easy to deduce the standard closed-graph and open-mapping theorems for Banach spaces from A.1. But this theorem does not (despite similarities) seem to afford an alternative proof of 3.5.2.

BIBLIOGRAPHY

J.C.Alexander

(1) Compact Banach algebras, Proc. London Math. Soc. (3) 18 (1968), 1-18.

T.Andô

(1) On fundamental properties of a Banach space with a cone, Pacific J. Math. 12 (1962), 1163-1169.

H.Bauer

(1) Über die Fortsetzung positiver Linearformen, Bayer. Akad. Wiss. Math.-Nat. Kl. S.-B. 1957 (1958), 177-190.

G.Birkhoff

(1) Lattice Theory, Amer. Math. Soc. Coll. Pub. 25, New York (1948).

E.Bishop and R.R.Phelps

(1) Support functionals of a convex set, Amer. Math. Soc. Proc. Symp. Pure Maths. VII, Convexity (1963), 27-35.

W.E.Bonnice and R.J.Silverman

(1) The Hahn-Banach extension and the least upper bound properties are equivalent, Proc. Amer. Math. Soc. 18 (1967), 843-849.

F.F.Bonsall

(1) Sublinear functionals and ideals in partially ordered vector spaces, Proc. London Math. Soc. (3) 4 (1954), 402-418.

(2) Regular ideals of partially ordered vector spaces, Proc. London Math. Soc. (3) 6 (1956), 626-640.

(3) Extreme maximal ideals of a partially ordered vector space, Proc. Amer. Math. Soc. 7 (1956), 831-837.

(4) Semi-algebras of continuous functions, Proc. London Math. Soc. (3) 10 (1960) 122-140.

(5) Locally compact semi-algebras, Proc. London Math. Soc. (3) 13 (1963), 51-70.

(6) Algebraic properties of some convex cones of functions, Quart. J. Math. Oxford (2) 14 (1963), 225-230.

(7) Compact linear operators from an algebraic standpoint, Glasgow Math. J. 8 (1967), 41-49.

F.F.Bonsall, J.Lindenstrauss and R.R.Phelps

(1) Extreme positive operators on algebras of functions, Math. Scand. 18 (1966), 161-182.

F.F.Bonsall and B.J.Tomiuk

(1) The semi-algebra generated by a compact linear operator, Proc. Edinburgh Math. Soc. (2) 14 (1965), 177-195.

G.A.Converse

(1) Extreme positive operators on C(X) which commute with given operators. Trans. Amer. Math. Soc. 138 (1969), 149-158.

E.B.Davies

(1) The structure and ideal theory of the pre-dual of a Banach lattice, Trans. Amer. Math. Soc. 131 (1968), 544-555.

M.M.Day

(1) Normed Linear Spaces, Springer, Berlin (1958; 2nd. ed. 1962).

D.A.Edwards

(1) On the homeomorphic affine embedding of a locally compact cone into a Banach dual space endowed with the vague topology, Proc. London Math. Soc. (3) 14 (1964), 399-414.

A.J.Ellis

(1) The duality of partially ordered normed linear spaces, J. London Math. Soc. 39 (1964), 730-744.

(2) Perfect order ideals, J.London Math. Soc. 40 (1965), 288-294.

(3) Linear operators in partially ordered normed vector spaces. J.London Math. Soc. 41 (1966), 323-332.

(4) Minimal decompositions in partially ordered vector spaces, Proc. Cambridge Phil. Soc. 64 (1968), 989-1000.

D.H.Fremlin

(1) Topological Riesz Spaces and Measure Theory, (to appear).

H.Freudenthal

(1) Teilweise geordnete Moduln, Proc. Akad. Wet. Amsterdam 39 (1936), 641-651.

L.Fuchs

(1) Partially Ordered Algebraic Systems, Pergamon, Oxford (1963).

(2) Riesz Vector Spaces and Riesz Algebras, Queen's Univ., Kingston, Ont. (1966).

H.Gordon

(1) Topologies and projections on Riesz spaces, Trans. Amer. Math. Soc. 94(1960), 529-551.

J.Grosberg and M.G.Krein

(1) Sur la décomposition des fonctionelles en composantes positives, Doklady Akad, Nauk SSSR 25 (1939), 723-726.

O.Hustad

(1) Convex cones with properties related to weak local compactness, Math. Scand. 11 (1962), 79-90.

G.J.O.Jameson

(1) Topological M-spaces, Math. Z. 103 (1968), 139-150.

(2) Allied subsets of topological groups and linear spaces, Proc. London Math. Soc. (3) 18 (1968), 653-690.

(3) Nearly directed subspaces of partially ordered linear spaces, Proc. Edinburgh Math. Soc. (2) 16 (1968), 135-144.

(4) Discrete and extremal positive linear mappings, J.London Math. Soc. 44 (1969), 559-564.

(5) Convex Series, Math. Z. (to appear).

M.A.Kaashoek and T.T.West

(1) Compact semi-groups in commutative Banach algebras, Proc. Cambridge Phil. Soc. 66 (1969), 265-274.

R.V.Kadison

(1) A representation theory for commutative topological algebras, Memoirs Amer. Math. Soc. no. 7, New York (1951).

S.Kakutani

(1) Concrete representation of abstract (L)-spaces and the mean ergodic theorem, Ann. of Math. 42 (1941), 523-537.

(2) Concrete representation of abstract (M)-spaces, Ann. of Math. 42 (1941), 994-1024.

L.Kantorovič

(1) Lineare halbgeordnete Räume, Mat. Sbornik 2 (44) (1937), 121-168.

J.L.Kelley

(1) General Topology, Van Nostrand, Princeton (1955).

J.L.Kelley and I.Namioka

(1) Linear Topological Spaces, Van Nostrand, Princeton (1963).

D.G.Kendall

(1) Simplexes and vector lattices, J. London Math. Soc. 37 (1962), 365-371.

J.Kist

(1) Indecomposable maximal ideals of a partially ordered vector space, J. London Math. Soc. 36 (1961), 436-438.

G.Köthe

(1) Topologische lineare Räume I, Springer, Berlin (1960).

M.G.Krein (see J.Grosberg)

M.G.Krein and M.A.Rutman

(1) Linear operators leaving invariant a cone in a Banach space (Russian), Uspehi Mat. Nauk 3 (1948), 3-95. English translation: Amer. Math. Soc. Translations (1) 10 (1962), 199-325.

J.Lindenstrauss (see F.F.Bonsall)

W.A.J.Luxemburg and A.C.Zaanen

(1) Riesz Spaces (Linear Vector Lattices), (to appear).

L.Nachbin

(1) Topology and Order, Van Nostrand, Princeton (1965).

H.Nakano

(1) Modulared Semi-Ordered Linear Spaces, Maruzen, Tokyo (1950).

I.Namioka (see also J.L.Kelley)

(1) Partially ordered linear topological spaces, Memoirs Amer. Math. Soc. no. 24, New York (1957).

Kung-fu Ng

(1) A representation theorem for partially ordered Banach algebras, Proc. Cambridge Phil. Soc. 64 (1968), 53-59.

(2) The duality of partially ordered Banach spaces, Proc. London Math. Soc. (3) 19 (1969), 269-288.

(3) Solid sets in ordered topological vector spaces, J. London Math. Soc. (to appear).

A.L.Peressini

(1) On topologies in ordered vector spaces, Math. Ann. 144 (1961) 199-223.

(2) Ordered Topological Vector Spaces, Harper and Row, New York (1967).

R.R.Phelps (see also E.Bishop, F.F.Bonsall)

(1) Lectures on Choquet's Theorem, Van Nostrand, Princeton (1966).

J.D.Pryce

(1) On type F semi-algebras of continuous functions, Quart. J. Math. Oxford (2) 16 (1965), 65-71.

(2) On the representation and some separation properties of semi-extremal subsets of convex sets, Quart. J. Maths. Oxford (2) 20 (1969), 367-382.

C.E.Rickart

(1) General Theory of Banach Algebras, Van Nostrand, Princeton (1960).

J.Riedl

(1) Partially ordered locally convex vector spaces and extensions of positive continuous linear mappings, Math. Ann. 157 (1964), 95-124.

F.Riesz

(1) Sur quelques notions fondamentales dans la théorie générale des operations linéaires, Ann. of Math. 41 (1940), 174-206.

G.T.Roberts

(1) Topologies in vector lattices, Proc. Cambridge Phil. Soc. 48 (1952), 533-546.

M.A.Rutman (see M.G.Krein)

H.H.Schaefer

(1) Halbgeordnete lokalkonvexe Vektorräume, Math. Ann. 135 (1958), 115-141.

(2) Halbgeordnete lokalkonvexe Vektorräume II, Math. Ann. 138 (1959), 259-286.

(3) Halbgeordnete lokalkonvexe Vektorräume III, Math. Ann. 141 (1960), 113-142.

(4) Topological Vector Spaces, Macmillan, New York (1966).

R.J.Silverman (see W.E.Bonnice)

G.F.Simmons

(1) Introduction to Topology and Modern Analysis, McGraw-Hill, New York (1963).

B.J.Tomiuk (see F.F.Bonsall)

B.Z.Vulih

(1) Introduction to the Theory of Partially Ordered Spaces (Russian), Fizmatgiz, Moscow (1961). English translation: Wolters-Noordhoff, Groningen (1967).

T.T.West (see M.A.Kaashoek).

J.D.Weston

(1) The decomposition of a continuous linear functional into non-negative components, Math. Scand. 5 (1957), 54-56.

Yau-chuen Wong

(1) Locally 0-convex Riesz spaces, Proc. London Math. Soc. (3) 19 (1969), 289-309.

(2) Order-infrabarrelled Riesz spaces, Math. Annalen 183 (1969), 17-32.

(3) The order-bound topology in Riesz spaces, Proc. Cambridge Phil. Soc. (to appear).

A.C.Zaanen (see W.A.J.Luxemburg).

LIST OF SYMBOLS

INDEX

Offsetdruck: Julius Beltz, Weinheim/Bergstr

Vol. 72: The Syntax and Semantics of Infinitary Languages. Edited by J. Barwise. IV, 268 pages. 1968. DM 18,– / $ 5.00

Vol. 73: P. E. Conner, Lectures on the Action of a Finite Group. IV, 123 pages. 1968. DM 10,– / $ 2.80

Vol. 74: A. Fröhlich, Formal Groups. IV, 140 pages. 1968. DM 12,– / $ 3.30

Vol. 75: G. Lumer, Algèbres de fonctions et espaces de Hardy. VI, 80 pages. 1968. DM 8,– / $ 2.20

Vol. 76: R. G. Swan, Algebraic K-Theory. IV, 262 pages. 1968. DM 18,– / $ 5.00

Vol. 77: P.-A. Meyer, Processus de Markov: la frontière de Martin. IV, 123 pages. 1968. DM 10,– / $ 2.80

Vol. 78: H. Herrlich, Topologische Reflexionen und Coreflexionen. XVI, 166 Seiten. 1968. DM 12,– / $ 3.30

Vol. 79: A. Grothendieck, Catégories Cofibrées Additives et Complexe Cotangent Relatif. IV, 167 pages. 1968. DM 12,– / $ 3.30

Vol. 80: Seminar on Triples and Categorical Homology Theory. Edited by B. Eckmann. IV, 398 pages. 1969. DM 20,– / $ 5.50

Vol. 81: J.-P. Eckmann et M. Guenin, Méthodes Algébriques en Mécanique Statistique. VI, 131 pages. 1969. DM 12,– / $ 3.30

Vol. 82: J. Wloka, Grundräume und verallgemeinerte Funktionen. VIII, 131 Seiten. 1969. DM 12,– / $ 3.30

Vol. 83: O. Zariski, An Introduction to the Theory of Algebraic Surfaces. IV, 100 pages. 1969. DM 8,– / $ 2.20

Vol. 84: H. Lüneburg, Transitive Erweiterungen endlicher Permutationsgruppen. IV, 119 Seiten. 1969. DM 10,– / $ 2.80

Vol. 85: P. Cartier et D. Foata, Problèmes combinatoires de commutation et réarrangements. IV, 88 pages. 1969. DM 8,– / $ 2.20

Vol. 86: Category Theory, Homology Theory and their Applications I. Edited by P. Hilton. VI, 216 pages. 1969. DM 16,– / $ 4.40

Vol. 87: M. Tierney, Categorical Constructions in Stable Homotopy Theory. IV, 65 pages. 1969. DM 6,– / $ 1.70

Vol. 88: Séminaire de Probabilités III. IV, 229 pages. 1969. DM 18,– / $ 5.00

Vol. 89: Probability and Information Theory. Edited by M. Behara, K. Krickeberg and J. Wolfowitz. IV, 256 pages. 1969. DM 18,– / $ 5.00

Vol. 90: N. P. Bhatia and O. Hajek, Local Semi-Dynamical Systems. II, 157 pages. 1969. DM 14,– / $ 3.90

Vol. 91: N. N. Janenko, Die Zwischenschrittmethode zur Lösung mehrdimensionaler Probleme der mathematischen Physik. VIII, 194 Seiten. 1969. DM 16,80 / $ 4.70

Vol. 92: Category Theory, Homology Theory and their Applications II. Edited by P. Hilton. V, 308 pages. 1969. DM 20,– / $ 5.50

Vol. 93: K. R. Parthasarathy, Multipliers on Locally Compact Groups. III, 54 pages. 1969. DM 5,60 / $ 1.60

Vol. 94: M. Machover and J. Hirschfeld, Lectures on Non-Standard Analysis. VI, 79 pages. 1969. DM 6,– / $ 1.70

Vol. 95: A. S. Troelstra, Principles of Intuitionism. II, 111 pages. 1969. DM 10,– / $ 2.80

Vol. 96: H.-B. Brinkmann und D. Puppe, Abelsche und exakte Kategorien, Korrespondenzen. V, 141 Seiten. 1969. DM 10,– / $ 2.80

Vol. 97: S. O. Chase and M. E. Sweedler, Hopf Algebras and Galois theory. II, 133 pages. 1969. DM 10,– / $ 2.80

Vol. 98: M. Heins, Hardy Classes on Riemann Surfaces. III, 106 pages. 1969. DM 10,– / $ 2.80

Vol. 99: Category Theory, Homology Theory and their Applications III. Edited by P. Hilton. IV, 489 pages. 1969. DM 24,– / $ 6.60

Vol. 100: M. Artin and B. Mazur, Etale Homotopy. II, 196 Seiten. 1969. DM 12,– / $ 3.30

Vol. 101: G. P. Szegö et G. Treccani, Semigruppi di Trasformazioni Multivoche. VI, 177 pages. 1969. DM 14,– / $ 3.90

Vol. 102: F. Stummel, Rand- und Eigenwertaufgaben in Sobolewschen Räumen. VIII, 386 Seiten. 1969. DM 20,– / $ 5.50

Vol. 103: Lectures in Modern Analysis and Applications I. Edited by C. T. Taam. VII, 162 pages. 1969. DM 12,– / $ 3.30

Vol. 104: G. H. Pimbley, Jr., Eigenfunction Branches of Nonlinear Operators and their Bifurcations. II, 128 pages. 1969. DM 10,– / $ 2.80

Vol. 105: R. Larsen, The Multiplier Problem. VII, 284 pages. 1969. DM 18,– / $ 5.00

Vol. 106: Reports of the Midwest Category Seminar III. Edited by S. Mac Lane. III, 247 pages. 1969. DM 16,– / $ 4.40

Vol. 107: A. Peyerimhoff, Lectures on Summability. III, 111 pages. 1969. DM 8,– / $ 2.20

Vol. 108: Algebraic K-Theory and its Geometric Applications. Edited by R. M. F. Moss and C. B. Thomas. IV, 86 pages. 1969. DM 6,– / $ 1.70

Vol. 109: Conference on the Numerical Solution of Differential Equ tions. Edited by J. Ll. Morris. VI, 275 pages. 1969. DM 18,– / $ 5.

Vol. 110: The Many Facets of Graph Theory. Edited by G. Chartra and S. F. Kapoor. VIII, 290 pages. 1969. DM 18,– / $ 5.00

Vol. 111: K. H. Mayer, Relationen zwischen charakteristischen Zahle III, 99 Seiten. 1969. DM 8,– / $ 2.20

Vol. 112: Colloquium on Methods of Optimization. Edited N. N. Moiseev. IV, 293 pages. 1970. DM 18,– / $ 5.00

Vol. 113: R. Wille, Kongruenzklassengeometrien. III, 99 Seiten. 19 DM 8,– / $ 2.20

Vol. 114: H. Jacquet and R. P. Langlands, Automorphic Forms on GL(VII, 548 pages. 1970. DM 24,– / $ 6.60

Vol. 115: K. H. Roggenkamp and V. Huber-Dyson, Lattices over Order XIX, 290 pages. 1970. DM 18,– / $ 5.00

Vol. 116: Séminaire Pierre Lelong (Analyse) Année 1969. IV, 195 pag 1970. DM 14,– / $ 3.90

Vol. 117: Y. Meyer, Nombres de Pisot, Nombres de Salem et Analy Harmonique. 63 pages. 1970. DM 6,– / $ 1.70

Vol. 118: Proceedings of the 15th Scandinavian Congress, Oslo 19 Edited by K. E. Aubert and W. Ljunggren. IV, 162 pages. 1970. DM 12,– / $ 3.30

Vol. 119: M. Raynaud, Faisceaux amples sur les schémas en group et les espaces homogènes. III, 219 pages. 1970. DM 14,– / $ 3.90

Vol. 120: D. Siefkes, Büchi's Monadic Second Order Success Arithmetic. XII, 130 Seiten. 1970. DM 12,– / $ 3.30

Vol. 121: H. S. Bear, Lectures on Gleason Parts. III, 47 pages. 19 DM 6,– / $ 1.70

Vol. 122: H. Zieschang, E. Vogt und H.-D. Coldewey, Flächen u ebene diskontinuierliche Gruppen. VIII, 203 Seiten. 1970. DM 16, $ 4.40

Vol. 123: A. V. Jategaonkar, Left Principal Ideal Rings. VI, 145 pag 1970. DM 12,– / $ 3.30

Vol. 124: Séminare de Probabilités IV. Edited by P. A. Meyer. IV, 2 pages. 1970. DM 20,– / $ 5.50

Vol. 125: Symposium on Automatic Demonstration. V, 310 pages. 19 DM 20,– / $ 5.50

Vol. 126: P. Schapira, Théorie des Hyperfonctions. XI, 157 pages. 19 DM 14,– / $ 3.90

Vol. 127: I. Stewart, Lie Algebras. IV, 97 pages. 1970. DM 10,– / $ 2

Vol. 128: M. Takesaki, Tomita's Theory of Modular Hilbert Algeb and its Applications. II, 123 pages. 1970. DM 10,– / $ 2.80

Vol. 129: K. H. Hofmann, The Duality of Compact Semigroups a C*-Bigebras. XII, 142 pages. 1970. DM 14,– / $ 3.90

Vol. 130: F. Lorenz, Quadratische Formen über Körpern. II, 77 Sei 1970. DM 8,– / $ 2.20

Vol. 131: A. Borel et al., Seminar on Algebraic Groups and Rela Finite Groups. VII, 321 pages. 1970. DM 22,– / $ 6.10

Vol. 132: Symposium on Optimization. III, 348 pages. 1970. DM 22, $ 6.10

Vol. 133: F. Topsøe, Topology and Measure. XIV, 79 pages. 19 DM 8,– / $ 2.20

Vol. 134: L. Smith, Lectures on the Eilenberg-Moore Spectral Sequen VII, 142 pages. 1970. DM 14,– / $ 3.90

Vol. 135: W. Stoll, Value Distribution of Holomorphic Maps into Co pact Complex Manifolds. II, 267 pages. 1970. DM 18,– / $

Vol. 136: M. Karoubi et al., Séminaire Heidelberg-Saarbrücken-St buorg sur la K-Théorie. IV, 264 pages. 1970. DM 18,– / $ 5.00

Vol. 137: Reports of the Midwest Category Seminar IV. Edited S. MacLane. III, 139 pages. 1970. DM 12,– / $ 3.30

Vol. 138: D. Foata et M. Schützenberger, Théorie Géométrique Polynômes Eulériens. V, 94 pages. 1970. DM 10,– / $ 2.80

Vol. 139: A. Badrikian, Séminaire sur les Fonctions Aléatoires Li aires et les Mesures Cylindriques. VII, 221 pages. 1970. DM 18, $ 5.00

Vol. 140: Lectures in Modern Analysis and Applications II. Edited C. T. Taam. VI, 119 pages. 1970. DM 10,– / $ 2.80

Vol. 141: G. Jameson, Ordered Linear Spaces. XV, 194 pages. 19 DM 16,– / $ 4.40